BIOSEPARATION ENGINEERING

Ajay Kumar
Dept. of Biotechnology
Institute of Biomedical Education and Research
Mangalayatan University
Aligarh (U.P.)
India

Abhishek Awasthi
Dept. of Science, Technology & Engg.
La Trobe University Melbourne
Victoria
Australia

TECHSAR™
NEW DELHI

Published and Printed by

TechSar Pvt. Ltd.
4435-36/7, Ansari Road, Daryaganj
New Delhi–110 002 (India)
E-mail: tech_info@techarworld.com
Website: www.techarworld.com

ISBN: 978-93-80026-08-4

Reprint 2025

BIOSEPARATION ENGINEERING

FOREWORD

The last two decades have witnessed a huge expansion in biotechnology education in India and abroad. As on today, this subject has become one of the favourite choice among the students seeking career in biology. A large number of academic institutions are offering undergraduate and post graduate degree programmes in biotechnology. Even this subject is one of the option for senior secondary students in some Secondary School Board examinations. The biotechnology curricula at undergraduate and postgraduate programmes have very good coverage of basic biochemistry, microbiology, molecular biology, immunology, cell and tissue culture technology and bioprocess technology. This enables the students to develop fairly good skills in bioresource and bioprocess development. Comparatively less weightage is for down stream processing or bio-product recovery from the cells or culture media. Moreover, reading materials or manuals on down stream processing targeting our undergraduate or postgraduate level students are not available in the market.

The present book 'Bioseperations Engineering' authored by the young budding biotechnologists Mr. Abhishek Awasthi, B.Tech and Mr. Ajay Kumar, B. Tech. (Abhishek Awasthi and Ajay Kumar), is an excellent attempt on their part to fill the gap as very few such titles are in the reach of students in the market. I congratulate the authors and highly appreciate their efforts in writing this book. This book has come at a time when most of us in biotechnology teaching are feeling that down stream processing should find more place in curriculum. It has covered most of the techniques of bioseperation so lucidly providing details of physical, chemical and biological principles involved and much needed strategies and applications. I am sure that this treatise will prove to be highly useful one not only to undergraduate and postgraduate students but to all those who are involved in biotechnology teaching and research.

Professor T. C. Bhalla
Department of Biotechnology
Himachal Pradesh University
Summer Hill, Shimla-171005
Phone/Fax: 0177-2832154, 2831948
Email: bhallaatc@rediffmail.com

PREFACE

Historically, biotechnology was an art involved in the production of wines, beers and cheeses. Nowadays it involves a series of advanced technologies spanning biology, chemistry and process engineering. Biotechnological processes have a three-component central core, in which one part is concerned with obtaining the best biological catalyst for a specific function or process, the second part creates, the best possible environment for the catalyst to perform, and the third part (downstream processing) is concerned with the separation and purification of an essential product or products from a fermentation process. This third aspect of biotechnology, namely, downstream processing or DSP, can be a technically difficult and expensive procedure, and is the least understood area of biotechnology. Bioseperations engineering, a comprehensive approach to down stream processing, is primarily concerned with initial separation of the bioreactor broth or medium into a liquid phase and a solid phase, and subsequent concentration and purification of the product. Processing will normally involve more than one stage and methods in use or proposed range from conventional to the arcane, including distillation, centrifuging, filtration, ultrafiltration, solvent extraction, absorption, selective membrane technology, reverse osmosis, molecular sieves, electrophoresis and chromatography. Final products of downstream purification stages should have some degree of stability, which is best achieved for most products by using some form of drying. In practice this is achieved by spray-drying, fluidized-bed drying or by freeze drying and the method of choice is product and cost dependent. Purity and stability are the hallmarks of the most high value biotechnological products like this edition of Bioseperations engineering. Simple language, lucid expression, in-depth concept explanation, and profusely illustrated diagrams will definitely make this book an erudition key for its reader.

Ajay Kumar
Abhishek Awasthi

ACKNOWLEDGEMENTS

We extend our sincere gratitude to Prof. Bhumitra Dev, Vice Chancellor and Prof. Ashok Kumar Dean, Institute of Biomedical Education and Research, Mangalayatan University, Beswan, Aligarh, for their constant inspiration in view of accomplishing the compilation embodied in this book. We are deeply indebted to Prof. Sanjay Mishra and Prof. S.K Maheshwari, Department of Biotechnology, CET, IFTM Campus, Moradabad for their intrinsic invaluable expert guidance and constant encouragement. Our thanks also go to our revered teacher, Dr. B.N. Mishra, Head, Department of Biotechnology, Institute of Engineering and Technology, Lucknow for generously extending the encouragement. We express our gratitude to Mrs. Shraddha Singh for her support and encouragement.

As we all know that no one is perfect in this world and this minute attempt of ours is to share our ideas among the Biotechnologists for enhancing the light of the lamp together. We shall welcome suggestions and critical comments from teachers and students and will incorporate the same with specific acknowledgements in future edition of this title.

Ajay Kumar
Abhishek Awasthi

CONTENTS

1

CHAPTER

BIOSEPARATION ENGINEERING: AN OVERVIEW

1.1 INTRODUCTION

One of the dominant characteristics of the high-technology industries is the overlap of skills needed between disciplines. The boundaries between fields of study are no longer sharp and distinct but rather are blurred to a considerable extent. Biotechnology is not an exception. Biotechnology and microbiology have been in use consciously or unconsciously by man since time immemorial for various purposes ranging from agriculture and poultry to medicine. The survival of man has depended on his ability to obtain extracts from nature for various uses in farming, poultry, medicine etc. However, the biomolecules or specimens of interest are almost always, never available in their pure form in nature.

In some cases, the concentration of the species of interest is too low while in some cases, the impurities present could hinder the target activity to be attained. These impurities could even prove to be toxic, which limits their efficiency. Thus, the need for removal of the unwanted materials from the sample/extract arises at this stage, and this process of isolation, extraction, concentration, purification and formulation of product or specimens of interest involves bioseparations and collectively is termed as bioseparations engineering. For products to move from the laboratory to commercial success requires more than just an understanding of the new biology; it requires an understanding of the fundamentals of engineering and their practice as applied to large-scale protein purification.

Each molecule in a mixture has its own identity in terms of various physical, chemical and electric properties such as size, weight, density, reactivity, charge etc: The various techniques employed in bioseparation make use of this fact to separate out the species of interest from the other molecules in the mixture. The techniques employed are also often chosen such that the end product can be recovered in its native form without any disruptions of any form.

In a small scale setup such as a laboratory, bioseparation is a relatively straight forward process without much of any hassles or hurdles. On the other hand, in a large scale setup such as that of an industry, with a daily turn over of more than a thousand liters per day, has very high complexity of the process. In addition, there is also the. The setup of facilities for such large scale concentration and purification, with the additional factor of adhering to the rules stipulated by the various regulatory boards, requires extensive research, leading to the development of a field called 'Bioseparation engineering'.

1.2 ADVENT OF BIOSEPARATION PROCESSES

Bioseparation in its basic form such as filtration, drying etc. has been exploited by man ever since the dawn of civilization. Over time as man gained deeper understanding of the underlying principles of biology and biochemistry, man began to identify the key organismal and molecular species present in extracts that brought about the effect under consideration. This understanding coupled with the technological advancement over time enabled man to exploit the various physical, chemical and electrical properties of the species of interest, both by improvisation of the already existing techniques as well as the development of more specialized techniques such as reverse osmosis, freeze drying etc. Thus finally enable the isolation of product with higher concentration and level of purity.

Water filtration for example is one such technique. A look into history shows that all major civilizations were clustered around water sources. The earliest records of water filtration date up to 2000 b.c. and the earliest recorded techniques indicate the use of crude sand or charcoal filters for the purpose. There are also records dating back to the 500 b.c indicating that the Greeks used cloth to filter water before drinking. As man kept realizing the importance of clean drinking water with time and identifying newer contaminants in water, newer more efficient filters with smaller pores and improved techniques were developed. Similar is the case with sedimentation and precipitation techniques, having existed in the basic form for over centuries and developing over time with advancements in knowledge and technology. Another such technique, freeze drying, has been in use by the ancient Peruvian Incas of the Andes, who use to store their food on mountain tops, in-order to preserve it. Freeze drying as we know of today was developed by the American scientist L.F. Shackell, in the early 20th century, from a lab equipment called the "chemical pump", by replacing the ethyl ether used by the equipment with vaccuum and realizing that the specimen required to be frozen before commencement of the process. However it was not until the mid 20th century that the technique began to be used on a large scale.

The first recorded centrifuge built for separation purposes was by Antonin Prandl from Germany in 1864 and a similar centrifuge designed to operate in a continuous fashion was later commercialized by Gustaf de Laval of Sweden. The ultracentrifuge was developed about 1920 by the Swedish chemist Theodor Svedberg and improved by the American physicist Jesse Beams. The era of Throdor Svedberg is considered to be the significant for the developments in centrifuge equipments and for the same this eminent scientist was awarded the Noble prize.

Most of the relatively specialized techniques in use today, such as freeze concentration, were developed in the 20th century. The Chromatography technique, for example, was invented in 1900 by Mikhail Semyonovich Tsvet, a Russian Botanist. The electrophoretic phenomenon of molecules was first discovered by a scientist Reuss from Russia. The first electrophoresis equipment for bioseparation purposes was invented by the American geneticist Oliver Smithies in the mid 19th century, for aiding his molecular studies.

Recent developments in the bioseparations engineering show immense potential in that, it might be possible to discard the RIPP (Recovery, Isolation, Purification and Polishing) scheme from Down Stream Processing. Membrane and monolith chromatography techniques, expanded chromatography, high resolution ultrafiltration and hybrid bioseparations are among the most promising of the emerging techniques. High throughput membrane processes can be fine tuned to give a high degree of selectivity.

1.3 WHAT IS BIOSEPARATIONS ENGINEERING?

Bioseparation engineering is a term used to denote the design, development and improvisation of apparatus and techniques employed for the concentration, purification and isolation of biomolecules or specimens of interest contained in a media. The process by which the molecules or specimen of interest is separated out from the media in the finished form ready for use or marketing is known as downstream processing (DSP). The concentration and purity levels demanded by the specialized requirements today, necessitates the coupling of more than one type of technique to deliver. This case is particularly true of industrial biotechnology where the powerful competition factor comes into play. In such cases, for the production control and management and the optimization of the whole process for maximum output in terms of quantity and quality, a scheme called the RIPP (Recovery, Isolation, Purification and Polishing) is followed. The idea behind the scheme is the use of lower resolution techniques, like filtration, first, for product recovery, followed by high resolution techniques, such as electrophoresis, for the purification and isolation of the product. Polishing refers to the processes involved in getting the purified/ isolated product into a form ready for packaging and marketing.

Bioseparation, directly or indirectly has an almost vital role in a common man's daily life today. In a general household, for example, tea is made using filter tea bags, safe drinking water is obtained through reverse osmosis filters and much more. In the case of the industrial sector, bioseparations engineering holds a central position, where the very design of the industrial manufacturing setup is largely influenced by the nature of the final product required, coupled with the energy considerations, time factor for each step in the process pipeline and the peripheral geometry, dictating the product concentration and purification apparatus setup, which further influences the production vessels. Also, the down-stream processing accounts for more than 40% of the production costs on a general scale.

1.4 COMMON DESIGNING STEPS FOR BIOSEPARATION PROCESS

One of the basic and most important things to realize while designing any process is that the design will not be better than the accuracy of data upon which it is based. At large scale it is a common practice to miss some of the basic data like phase equilibria and physical properties, in order to increase speed of tasks. The steps below are the common steps for designing of all separation processes and all of them are vital.

1. Establishment of bases.
   a. Composition of feed and product(s).
   b. Rate to be processed.
   c. Operational conditions like pressure and temperature.
   d. Special conditions like presence of suspended particles, excursions of pH, temperature, or pressure.
2. Obtain basic data.
   a. Density, viscosity, diffusion coefficients.
   b. Phase equilibria or flux data for membranes.
   c. Efficiency or mass transfer data. (Basically the plant data)
3. Perform process selection.
   a. Critically influenced by bases and basic data.
   b. Complete performance and economic evaluations.
4. Complete process design.
   a. Verify economics.
   b. May need to change process selection.

1.5 CLASSIFICATION AND NATURE OF BIOSEPARATIONS

The production of biologicals is basically divided into two categories, which are upstream and downstream operations. The term upstream refers to the production of the desired compound or molecule in some raw form, which includes fermentation and cell culture. While downstream part includes the recovery and purification of this product. The upstream and downstream are interdependent to each other and the choices made in the design of the upstream side affect the design of downstream side.

Integration of upstream and downstream operations often results in a product with lower cost and higher quality. The most skillful improvement of a downstream recovery process will produce at best a tenfold increase in the yield. However, skillful improvement of the upstream fermentation process through selection of high-producing organisms, high-growth media, and optimal operating conditions may produce a thousand-fold increase in production.

It is advisable to consider the impact of improvements in the upstream and downstream sides of the process. An improvement in the downstream recovery process can result up to ten fold increase in yield, while if the improvement is at upstream end through selection of high-producing organisms, high-growth media, and optimal operating conditions may produce a thousand-fold increase in the production.

The production process in biotechnology industry consists of many steps involving the cultivation of cells followed by the product manufacture. The former known as upstream processing (USP) is carried out in bioreactors and the latter known as downstream processing (DSP) in production reactors; forming an overall two phase production process.

The key differences between upstream and downstream processes is the scale of operation. Some of the differences are between upstream and downstream operations are tabulated below:

Upstream	***Downstream***
Live organisms	No organisms
Large volumes	Progressively smaller volumes
Many components	Progressively fewer components
Heterogeneous mixture	Homogeneous mixture
Capital intensive	Labor intensive
Automated	Manual/ Automated
Fewer operations	Many operations
Lower cost materials	Higher cost materials

1.6 UPSTREAM PROCESSING

For any operation the upstream and downstream process interaction happens superficially at the point of transfer. The final product of the upstream operation is the raw material for the downstream operation. The upstream group is assigned to produce the product in a crude form, while downstream must remove the impurities, including those added at upstream stages.

The USP is concerned with processes involved in the culture of the cells required for the product manufacture, such as buffer and media preparation, and those involved in their extraction, such as chromatography etc. The cells are initially streaked on agar or other suitable media, from where colonies are picked and transferred to a test tube containing two milliliters of suitable pre-prepared culture media and grown for an appropriate time before separating the cells and transferring them to a flask containing twenty milliliters of the culture media. Thereafter the process is repeated, transferring the cells to two liters and twenty liters of bioreactors successively after providing sufficient time for the cells to grow and multiply to the required extent. This phase requires vessels for media and buffer preparation, supply and transfer as well as apparatus for the cell separation such as chromatography columns, filters etc.

The separated cells are then transferred to the production reactors fed with the culture media containing nutrients and other materials boosting the production of the required metabolite and then left to metabolize (fermentation) for a sufficient time period, which in most cases is forty-eight hours. The production reactor at this stage is maintained at stipulated conditions of temperature, pressure and dissolved oxygen, which again in most cases is 35°C, 70 PSI and 60 DO.

There are four variable factors in USP that affect the DSP: namely, the volume of material delivered, the amount of impurities relative to the product, the nature of the impurities, and the phases present (solid, liquid, or gas).

1.7 DOWNSTREAM PROCESSING

To begin our designing of the large scale purification process, our starting place is the ending place for the fermentation step. It is safe to say that the design of a large scale purification process is an evolutionary method that relies on an existing process, a set of general rules, and some degree of expertise. Figure 1 describes a general downstream process flowchart.

The customary definition of downstream includes everything after product synthesis and consequently the process begins with the recovery operation. A lot of variation has been observed in the stages of purification processes, but in general it starts with removal of insoluble material followed by isolation, purification of the product and finishes with its polishing.

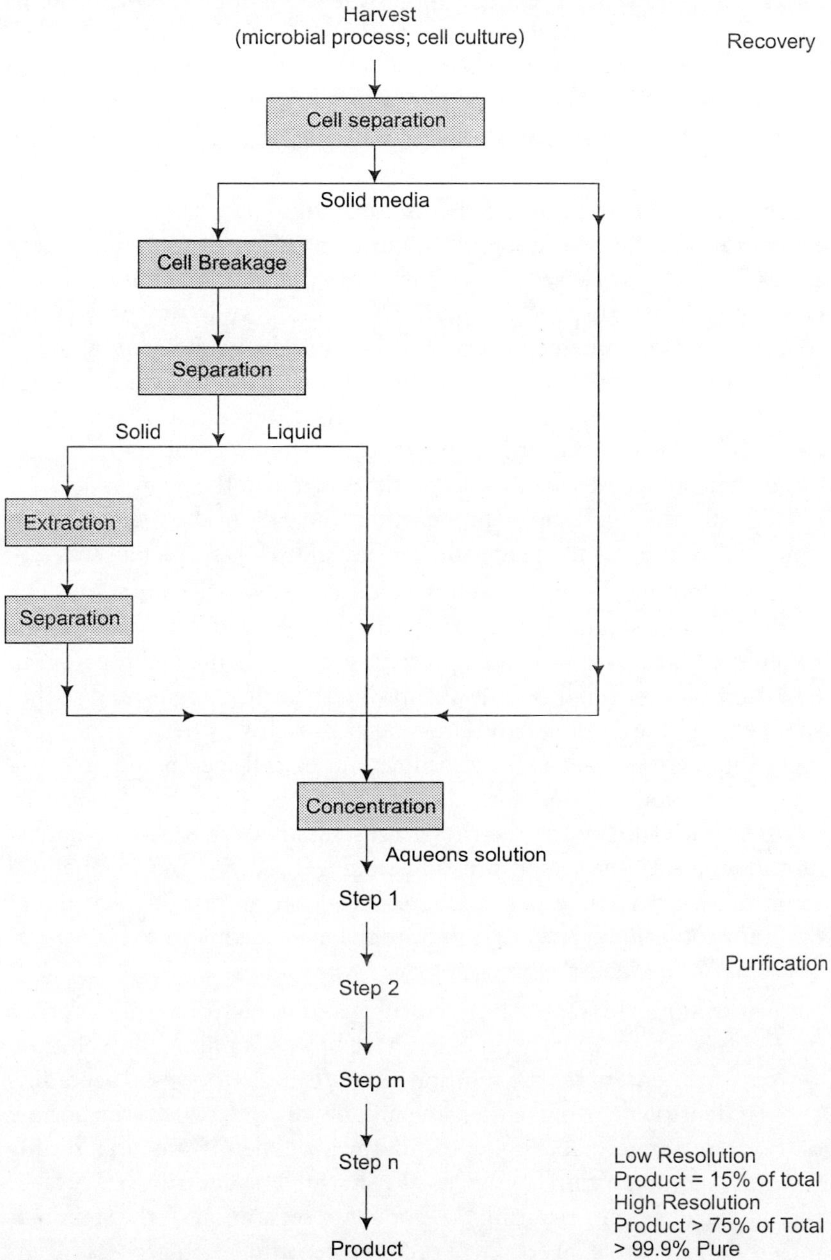

Fig. 1 Flowchart of downstream processes

All processes following the completion of the fermentation process, up to the dispatch of the product for marketing falls under the DSP. There are several steps involved in this phase, the order of which are, in compliance with the RIPP scheme, are as described below. The steps are the same in most of the industries, with only a few exceptions in terms of the number and order of the steps involved.

1. Filtration: In this step, the cells and other solid material present in the fermentation solution are separated out using filters of appropriate pore sizes. The process is based on the passage of the solution containing the dissolved solutes, through the pores of the filter, while the materials in solution with a diameter larger than the pore diameter are retained at the feed side of the filter. In addition, those particles which cannot traverse through the tortuous channels of the filter also get caught up in the membrane. As the filtration proceeds, the pores of the filter get clogged due to all the retention of particles in and at the membrane. This necessitates the use of appropriate setup to clean up the membrane on timely basis in order to ensure the continued use of the membrane.
2. Cell Disruption: Product recovery often requires cell disruption. The product may be soluble but found within the cell, or the product may be insoluble. Many over produced proteins are found clumped together in inclusion bodies, small nodules of insoluble protein segregated within the cell. These soluble and insoluble proteins, whether they are inclusion bodies found in yeast or bacteria or cell-membrane proteins, must be separated from the cellular components before they can be purified. Cell disruption methods are classified as non-mechanical methods, ultrasonics, high pressure homogenization, and mechanical grinding. Each of these has characteristics that make it favored under different conditions. The latter two are the most amenable to large-scale purification in modern biotechnology.
3. Centrifugation: Here the solution is subjected to high centripetal forces, by rotating at appropriate speeds, a vessel containing the solution, whereby the contents of the solution separate out in the vessel according to their densities, the higher density material resting towards the bottom of the vessel content at a higher speed, the lower density material resting towards the top portion of the vessel content. The centrifuge is rotated for a sufficient time, ensuring the complete separation of the materials into their respective layers. Generally the centrifugation process results in the formation of a supernatant fluid, at the top and one or more layers of precipitate. Depending on the 'dissolution properties' of the species of interest, the supernatant of the appropriate precipitate layer is collected for further processing.

4. Precipitation: Precipitation is the formation of insoluble solids in solution. This step is used to separate out the solute molecules from their solute phase by the addition of inorganic salts (salting out) or by decreasing the salt concentration (salting in), organic solvents, to change the dielectric properties of the solution leading to precipitation, or supersaturating the solution by the addition of compounds. The solids are then separated out by settling and sedimentation or centrifugation. The precipitation process is generally carried out at low temperatures. The main advantage of the precipitation process is that it helps in dealing with large volumes of solutions.
5. Extraction/Adsorption: Extraction can be used to either separate out a solute in solution, also called as leaching, or a liquid from a liquid-liquid solution, also called partitioning. The former is used very rarely in biotechnology, while the latter is applied routinely in separation processes. This process is based on the different solubility of a particular liquid or solid in different solvents, leading to the 'extraction' of the solute from one solvent to the other. A solvent immiscible with the original solvent, and having increased affinity for the solute as compared to the original solvent, is added to the solution with mixing and allowed to settle whereby the solute gets extracted out of the original solution into the added solvent. The two immiscible liquid phases are then separated out using distillation or other suitable processes. The added solvent is now obtained with the solute and is called the raffinate and the original solvent obtained after the extraction has been performed is called the extract. The extraction procedure is generally performed more than once to enable maximal recovery of the solute from the original solvent.

   Adsorption involves the selective collection of one or more components of the solute molecules in solution, at the surface of a specially selected microporous solid, like silica etc:, called adsorbant. This may be due to physical molecular forces, electrostatic forces or simply due to normal physical properties such as molecular size etc: The collected molecules present on the solid surface are called as adsorbates. The adsorbant can then be 'regenerated' easily by raising its temperature or reducing the concentration of the adsorbate, leading to the desorption of the adsorbates from the adsorbant making the adsorbant ready for re-use. More than one round of adsorption might be required depending the surface area and strength of the adsorbant used.
6. Membrane separation: The basic mechanism of action here is essentially the same as that of filtration except that the size of the particles being separated out here are much smaller compared to those in filtration.

This is in turn the effect of the pore size of the membrane used. Membranes used at this step of the separation are called semi-permeable membranes owing to the high degree selectivity of the molecules the membranes let pass through. One particular variation, routinely used is the 'Reverse osmosis' process, whereby pressure is applied on the side of the membrane containing the solution so as to facilitate the reverse process of osmosis, whereby the solvent molecules move across the membrane, resulting in the retention of a highly concentrated solution.

7. Chromatography: Chromatography involves the passing of a 'mobile phase', which in this case would be the solution/mixture containing the required product to be separated, through a 'stationary phase', which is usually a column of gel or matrix of suitable material. The stationary phase interact with the components of the mobile phase as it passes through the stationary phase, each component of the mobile phase passing through the column at different rates, depending on the strength of the interaction. The unbound fraction of molecules in the mixture/ solution come out with the effluent and the recovery of the bound fractions, often requires the column to be washed down with an elution buffer. In other cases, the stationary phase acts like as a filter to separate the components of the mobile phase based on the molecular size. The high precision of the technique enabling its use for the separation of proteins differing by even single amino acids, alongside the wide range of bases and operable conditions in which it can be used, makes it a valuable asset in industrial biotechnology DSP. This step marks the end of DSP for liquid products.
8. Electrophoresis: The solution is then subjected to electrophoresis, whereby the components of the solution are further separated, this time based on the electric properties of the components in solution. The solution/mixture is loaded onto a gel or a suitable conductive media of similar properties and subjected to an electric field, whereby the contents of the sample get separated out into their individual components. This process is based on the fact that, any molecule dropped into an conductive media, aquires a net surface charge, and on subjecting to an electric field in the media, migrates through the media, towards the oppositely charged electrode at a rate dependent on the charge magnitude, size of the molecule and the media properties such as, viscoscity in the case of liquid media, pore size in the case of gels and so on. The individual components eventually run out of the media and can be collected easily for further purification and conceentration. The product can be considered to be 'pure' at the end of this stage.

9. Drying: As the name suggests, this involves the removal of the residual content from the purified product by vaporization, to result in the formation of tablet and powder, the latter containing minimal moisture content as compared to the former. The process involves the heating of the product using contact method, where a heated surface is brought into contact with the wet solid where vaporization begins within the solid, diffusing to the surface, or the air dry method, where it is subjected to heated air, which causes the evaporation of water from the surface. In the case of heat labile product recovery, vaccum drying method is used, where an artificial vaccum, at room temperature, is created leading to the vaporization of water content of the solid. In the case of the tablet preparation, the tablets are further processed and pressed into the tablet form using one of several protocols. For powder preparation, the drying process is continued further until the maximal moisture content that can be separated from the solid has been removed. In most cases, the solid products are now ready for packaging and marketing.
10. Crystallization: In the case of materials like the common salt (NaCl), sugar and other inorganic salts, the solution is brought to a saturated state at elevated temperatures followed by force cooling, which forces the solid to precipitate out of solution. The crystals can be separated by techniques like filtration etc: The mother liquor is then sent back again for further concentration and purification followed by crystallization and the process continues up until the maximal product has been recovered from the solution.
11. Finished good formulation and packaging.

1.8 BASIC CONCEPT EXPLAINED

The above mentioned procedure is in the case of an extracellular product such as an extracellular enzymes released by yeast cells upon fermentation. If the product is intracellular, then in addition to the above mentioned steps, the product recovery procedure includes cell disruption as the first step in DSP, whereby the cells in the fermemtation media are subjected to extreme conditions such as ultrasonication, high pressure and mechanical homogenization, inorder to break open the cells and release the cell lysate containing the product of choice into the media. From here, the normal procedure as described above is followed, the cellular debris getting separated out at the filtration level.

Several specialized variations exist for each of the above mentioned techniques, such as ultrafiltration in the case of filtration and membrane filters, the use of filter vessels in the case of centrifuge, that enable the components of the media to filter out with the subjected centrifugal force as the centrifugation proceeds, freeze drying in the case of drying where a sudden drop of temperature under appropriate conditions followed by a reduction in pressure results in direct sublimation of the ice crystals formed, which is not only efficient in terms of the moisture separation but is much gentler on the products as compared to the other drying techniques and so on for the other techniques as well. Each technique with its major modifications will be existed in detail in the coming chapters.

Analysis: There will be several reactors and vessels operating simultaneously at any point along the process and a proper planning in terms of the time required for each step, the energy and material input and output etc: at each step etc: is essential to ensure the smooth working of the process pipeline. At the completion of every single batch of the process, the whole plant is treated and sterilized using superheated steam.

1.9 A FEW CASE STUDIES

Production of β-Amylase: A large sector of the industrial biotechnology is devoted to the production of synthetic enzymes in specifically selected microorganisms as compared to their extraction from naturally occurring sources. Specially selected strains of A. *Niger* are cultured separately and an inoculum from this culture is fed into the fermentation medium. The fermentation medium is prepared by the mixing of nutrients required for the cell growth, water and other additives required for production of β-Amylase by A. *Niger* cells, and then sterilized, following which the medium is inoculated with the cells, and is fed into a fermentor and then maintained at conditions favoring fermentation. After time has been provided to ensure sufficient levels of fermentation, the medium is subjected to broth separation techniques whereby the mycelium is removed. This is followed by concentration steps where a large amount of the water content is separated out from the solution. This concentrated solution is then treated with corn starch and iso-propanol for precipitation of other unwanted components of the solution. Thereafter, the roughly purified form of the product now obtained is used in several solid separation techniques to isolate out the enzyme from solution. The alchohol content, from the precipitation step as well as the fermentation reaction, is separated out from the product in these steps and fed back along the pipeline for use in the precipitation steps, for optimizing the production costs. The solid form of β-Amylase obtained now is a wet solid. This wet solid is then subjected to drying techniques whereby the moisture content of the solid, and trace amounts of alchohol that persisted from the previous separation steps, vaporize out from the solid, thus giving the final product as β-Amylase on corn starch. (see Figure 2.)

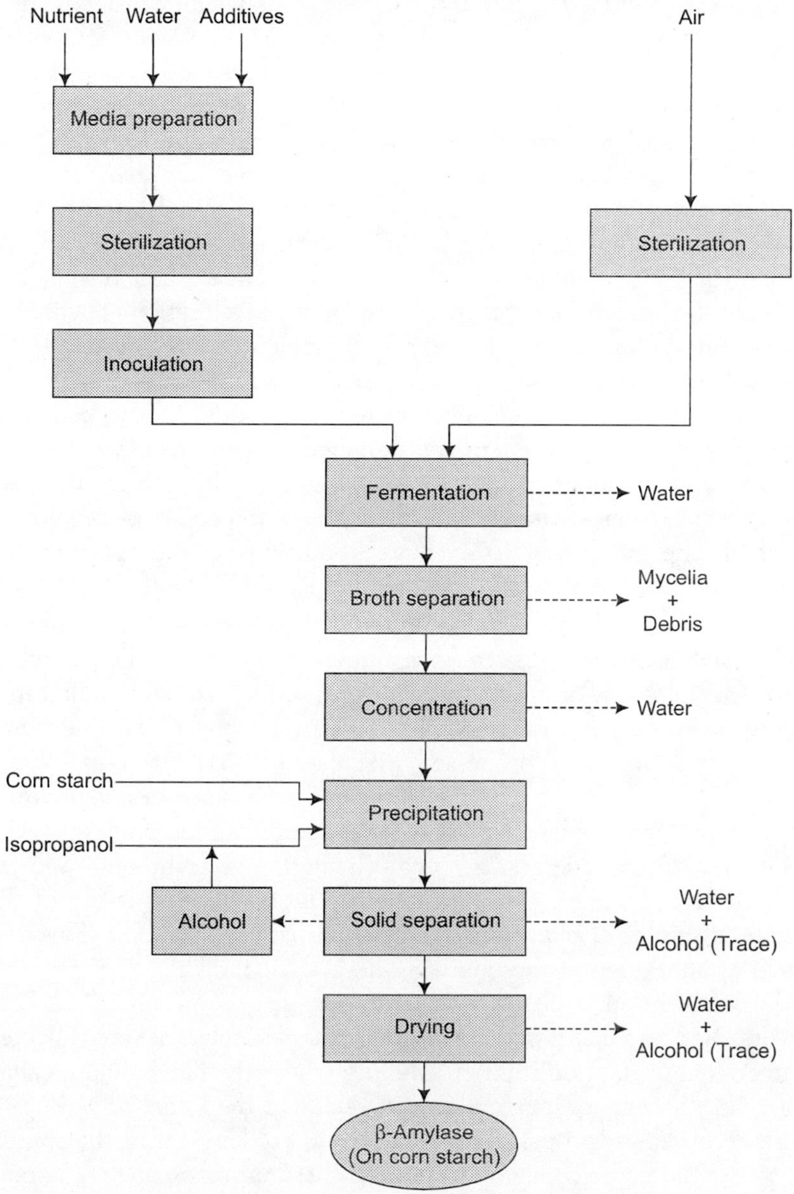

Fig. 2 Process synthesis flowchart for β-Amylase

Production of Penicillin: Among all the biologically produced antibiotics available in the market mostly moulds are used with only a few exceptions where bacteria are used. One particular example of antibiotic production using moulds is that of penicillin production. As in the case with β-Amylase, the microorganisms are grown separately and fed to a sterilized fermentation medium in a fermentor and provided sufficient time under suitable conditions to produce the product. The broth, including mycelium, from the fermentation media is then separated out as in the previous. Now, the penicillin in solution is separated out from the fermentation media by repeated cycles of solvent extraction, feeding the raffinate back into the extraction for maximal penicillin recovery. This extracted penicillin in solution is then concentrated by adsorption using activated carbon and then purified using solids separation, the solvent recovered in this process being fed back into the solvent extraction step. The purified solid form of the product is then treated for crystallization of penicillin using potassium/sodium acetate and the solids separated out and dried, the recovered solvent again being fed back into the solvent extraction step or discarded if containing excessive concentrations of impurities. Thus penicillin is purified and isolated from the fermentation media. Note, the absence of the use of separation techniques using excessive conditions like high temperature etc. This is because of the relatively delicate nature of the nature of antibiotics, due to which the subjection to extreme conditions could result in the denaturation of the antibiotics, thus degrading the product quality.

Production of Ethanol: Ethanol production on the industrial scale, is generally carried out using *Saccharomyces cerevisiae* acting on molasses. When starch is used instead as the source of sugars, there is a pre-initiation treatment of starch with enzymes necessary, to convert starch to fermentable sugars. The nutrients, molasses, additives and water are mixed, inoculated with *S. cerevisiae* and fed into thc fermentor for fermentation at stipulated conditions of temperature, pressure, pH etc: for a stipulated time. After the fermentation process is over, the media is made to undergo fractional distillation, a technique used for separating the components of a homogenous mixture by making use of the different boiling points of each component, by partial vaporization followed by progressive condensation of the mixture whereby the individual components condense out of the vapor state separate from each other and can be collected accordingly. The components of the mixture in this particular case are collected using a fractional distillation train of columns, where the first column, called the primary distillation column, collects the cell and the initial solvent. The vapors are then allowed to pass over to a second column, called the splitter column, where the vapor components get separated into two, one containing the butanol vapors and the other containing acetone/ethanol vapors. The former is recovered from the vapors by condensation in butanol columns. The latter of the two is first passed through an acetone column, where acetone is condensed and separated out from the vapors and the left over is passed over to the ethanol recovery column, where ethanol water mixtures of 55 to 98% concentrations are separated out in similar fashion.

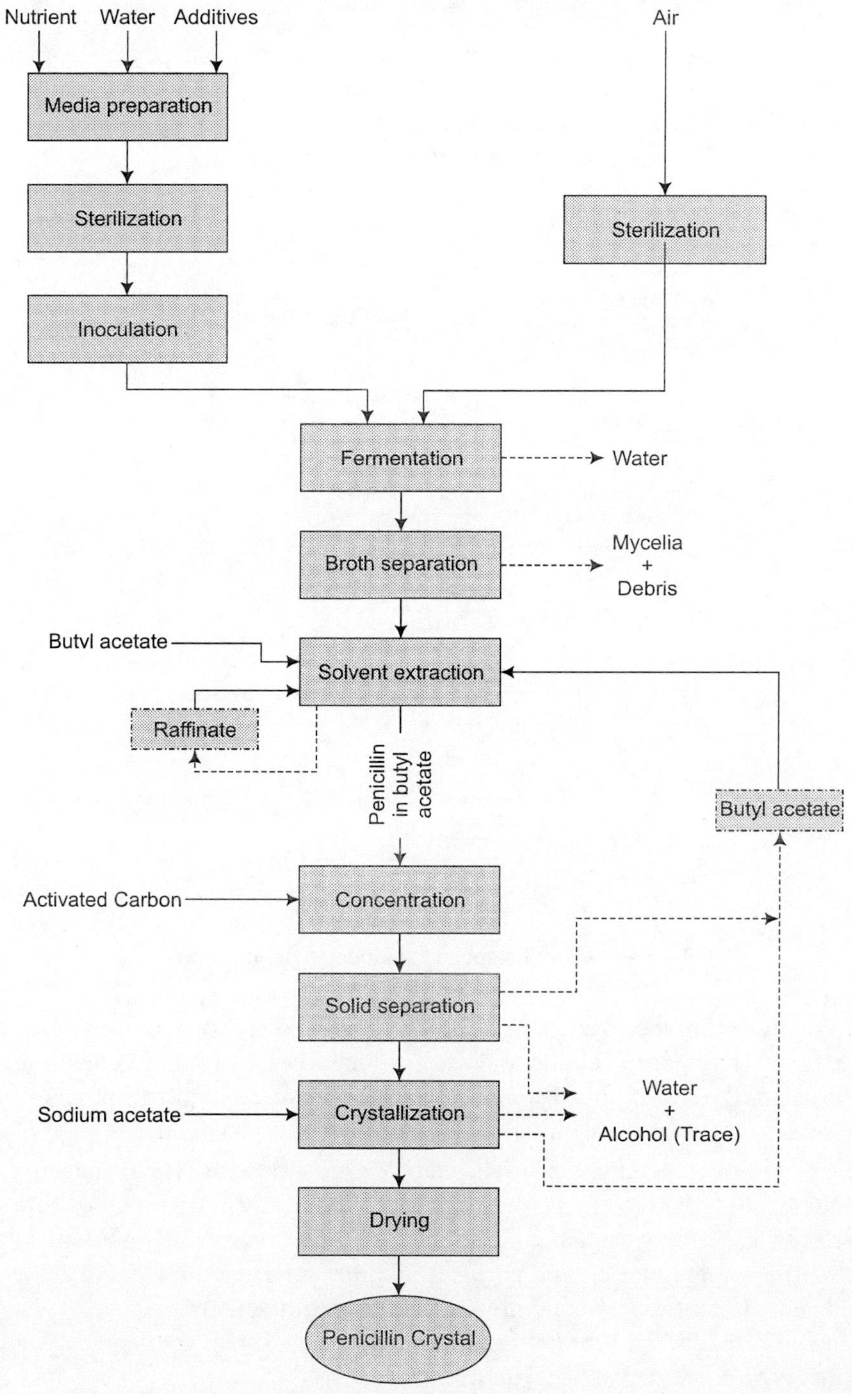

Fig. 3 Flowchart for the isolation and purification of Penicillin crystal

1.10 TEMPLATE ISOLATION AND PURIFICATION SCHEME

The degree of purity of any commercial biological product depends on the application. For example, in case of commercial enzymes the purity level ranges from raw enzymes to highly purified forms. Raw materials for the isolation of enzymes are animal organs, plant material, and microorganisms.

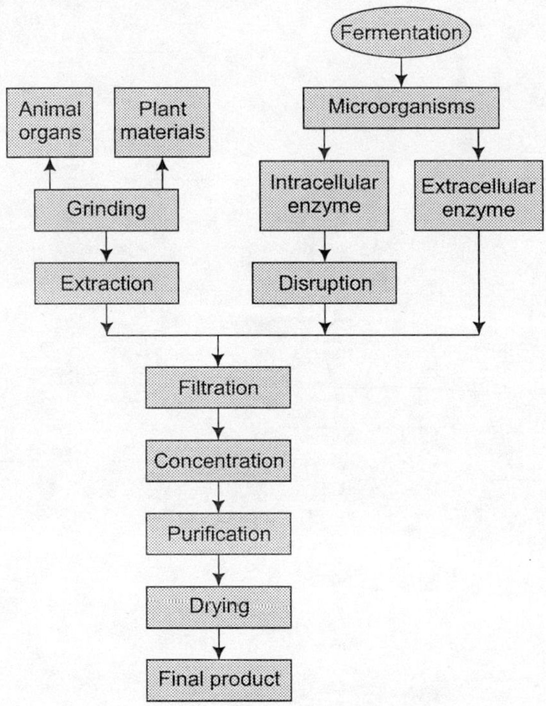

Fig. 4 Sequence of steps in the isolation of enzymes

Procedure for the enzyme purification will solely depend on their localization. Among intracellular and extracellular, the isolation of former one often involves the separation of complex biological mixtures while the latter are generally released into the medium with only a few other components.

Enzymes or the biocatalysts are the complex proteins with very high degree of specificity. In order to achieve that, they need to be only in their native state and due to this we only employ mild, specific methods for enzyme isolation. To optimize the whole isolation and purification process for enzymes, major areas viewed to maintain its native conformation are specific conditions of: (i) pH, (ii) temperature, and (iii) ionic strength. Figure below shows the sequence of steps involved in recovery of enzymes.

2

CHAPTER

CELL DISRUPTION

2.1 INTRODUCTION

Cell disruption is a sensitive process that is used to break the cell envelope, releasing all the intracellular components in the surrounding medium. Cell disruption focuses on obtaining the desired product from within the cell, and it is the cell wall that must be disrupted to allow the contents of the cell to flow out. The cell wall conveys its strength to the cell and can be formed from differing kinds of complex polysaccharides which are generally cross-linked by peptides to a degree and give different organisms varying levels of strength.

In essence, the objectives of cell disruption are as follows:

1. To solubilise the maximum amount of the product present in the cell whilst still maintaining maximum biological activity.
2. To avoid secondary alteration of the product that will render it useless e.g. denaturisation and oxidation.
3. To limit the detrimental effects of the disruption stage on the following separation steps.

The disruption of cells is an important stage in the isolation and manufacture of intracellular products for biopharmaceutical, food, nutraceutical, and agrochemical industries. From research level to production, many areas of biotechnology, particularly recombinant technology, necessitate the use of efficient, temperature controlled, and repeatable cell disruption systems.

The production of biologically interesting molecules can be done using cloning and culturing methods and also biological products are secreted from the cell or released during autolysis, the preparation of many others requires cell disruption to release intracellular material. There are a few factors that must be considered before cell disruption, like the volume or sample size of the cells to be disrupted. If only a few microliters of sample are available, care must be taken to minimize loss and to avoid cross contamination, but when hundreds or several thousands of liters of material are being processed then throughput, efficiency, and reproducibility are the important factors to be considered. The ease with which cell can be disrupted depends upon the cell structure and composition of its cell wall. As the difficulty of disruption increases (e.g. *E.coli*), more force is required to efficiently disrupt the cell. For even more difficult samples like yeast, there is a parallel increase in the processor power and cost. The most difficult samples (spores) require mechanical forces combined with chemical or enzymatic efforts. Over-disruption may also impact the desired product. For production scale processes, the time to disrupt the cells and reproducibility of the method becomes more important.

General problems associated with cell disruption include the liberation of DNA, which may increase the viscosity of the suspension and thus creates problem during separation processes. This may also affect further processing, such as pumping the suspension on to the next unit process and flow through chromatography columns. If mechanical disruption is used then heat is invariably generated, which denatures proteins unless appropriate cooling measures are implemented. Products released from eukaryotic cells are often subjected to degradation by hydrolytic enzymes liberated from disrupted lysosomes, however this damage can be reduced by the addition of enzyme inhibitors. Also cooling the cell extract and rapid processing may help to reduce the damage; moreover attempts may be made to produce mutant strains of the producer microorganism lacking the damaging enzymes.

In general, the cell disruption method is closely matched with the material that is desired from the cell studies. It is usually necessary to establish the minimum force of the disruption method that will yield the best product. It is rare that a cell disruption process produces a directly usable material, so subsequent purification events are necessary. Additionally, once the cells are disrupted, it is important to consider what components are present in the disruption media so that efficient purification is not impeded and it is also essential to protect the desired product from normal biological processes and from oxidation or other chemical events.

2.2 REVIEW OF LITERATURE

Micro-organisms are protected by extremely tough cell walls. In order to release their cellular contents a number of methods have been developed. Any potential method of disruption must ensure that labile materials are not denatured by the process or hydrolysed by enzymes present in the cell. The desired product can be obtained with minimum contamination by using a combination of different techniques to release product from specific locations from the cell. Cultured cells are normally easy to disrupt. Cells grown in suspension are collected by centrifugation, washed and resuspended in a lysis solution. Lysis is made complete by immediate vortexing or vigorous pipetting of the solution. Attached cells can be lysed directly on the culture plate.

Saccharomyces cerevisiae, *Pichia pastoris* and *E.coli* have long been used in the biopharmaceutical process as each has reasonably well known genetic systems and act as a good host to produce desired intracellular material such as proteins.

Animal tissues and cells: Most animal tissues can be processed fresh (unfrozen). It is important to keep fresh tissue cold and to process it quickly (within 30 minutes) after dissection. When disrupting fresh tissue, the cells need to be sheared immediately at the time the GITC lysis solution is added. This can be done by dispensing the lysis solution in the tube, adding the tissue and immediately sonicating or homogenizing. Samples should never be left sitting in lysis solution, undisrupted. Hard tissues such as bone, teeth and some hard tumors may require milling. Animal tissues that have been frozen after collection are disrupted by grinding in liquid nitrogen with a mortar and pestle. During this process, it is important that the equipment and tissue remain at cryogenic temperatures. The tissue should be dry and powdery after grinding. Grinding should be followed by thorough homogenization with a mechanical homogenizer in a lysis buffer. Cultured cells are normally easy to disrupt. Cells grown in suspension are collected by centrifugation, washed and resuspended in a lysis solution. Lysis is made complete by immediate vortexing or vigorous pipetting of the solution. Attached cells can be lysed directly on the culture plate. Lysis solution is added directly to the plate or flask and cells are scraped into the solution. The lysate is then transferred to a tube and vortexed or pipetted to ensure complete cellular disruption. Alternatively, cells can be detached, collected, rinsed with PBS to remove culture medium, and then lysed by vortexing or sonicating in the lysis solution. Placement of the flask or plate on ice while washing and lysing the cells will further protect the RNA from endogenous RNases released during the disruption process.

Plant tissues: Soft, fresh plant tissues from plants such as Nicotiana and Arabidopsis can often be disrupted by homogenization in lysis buffer alone. Other plant tissues, like pine needles, need to be ground dry, without liquid nitrogen. Some hard, woody plant materials may require freezing and grinding in liquid nitrogen or milling. Plant cell suspension cultures and calluses can be lysed by sonication in a lysis buffer for 0.5–2 minutes. The diversity of plants and plant tissue make it impossible to give a single recommendation for all.

However, most plant tissues typically contain polysaccharides and polyphenols that can coprecipitate with RNA and inhibit downstream assays. Treating a plant tissue lysate with polyvinylpyrrolidone (PVP) will precipitate such problematic components from the lysate before the actual RNA isolation is carried out.

Bacteria: Bacteria, like plants, are extremely diverse; therefore, it is difficult to make one recommendation for all bacteria. Bead milling will lyse most Gram positive and Gram negative bacteria, including mycobacterium. It can be performed by adding glass beads and lysis solution to a bacterial cell pellet and milling for a few minutes. It is possible to lyse some Gram negative bacteria by sonication in lysis solution alone. Bacterial cell walls can be digested with lysozyme to form spheroplasts. Gram positive bacteria usually require more rigorous digestion (increased incubation time, increased incubation temperature, etc.) than Gram negative organisms. The spheroplasts are then easily lysed with vigorous vortexing or sonication in lysis buffer.

E.coli cells were found to be sensitive to lysis by the supernatants of a variety of protease-positive Bacillus species when treated at 45°C but not when treated at 37°C. Different *E. coli* strains manifested different lysis sensitivities when treated at 45°C. When the lysis rates of *E. coli* cells at various stages of growth were investigated, post-exponential-phase cells were shown to be most sensitive to lysis. The lysis rate was inversely related to cell viability, and susceptibility appeared to be at least partly due to lysis of dead *E. coli* cells.

The recovery of intracellular protein from *E. coli* can also be carried by bead mill disintegration. Several parameters are considered: operating time, amount and size of beads, cellular suspension concentration, and presence of toluene and lysozyme.

Yeast and fungi: Yeast can be extremely difficult to disrupt because their cell walls may form capsules or nearly indestructible spores. There are several ways to approach yeast cell disruption. One of the most common and probably the most straightforward methods is mechanical disruption using a bead mill. Bead mills vigorously agitate a tube containing the sample, lysis buffer and small glass beads (0.5–1.0 mm). Alternatively, yeast cell walls can be digested with zymolase, glucalase and/or lyticase to produce spheroplasts that are readily lysed by vortexing in a guanidinium-based lysis buffer. Some specialized isolation methods for yeast exist which use such methods as boiling SDS or boiling phenol treatment to insure complete cell lysis.

To disrupt filamentous fungi, scrape the mycelial mat into a cold mortar, add liquid nitrogen and grind to a fine powder with a pestle. The powder can then be thoroughly homogenized or sonicated in lysis buffer to solubilize completely. As fungi may also be rich in polysaccharides, treatment with PVP may be helpful here too.

Fig. 1 Cell disruption of Yeast cells

2.3 PRINCIPLE AND THEORY

Cell disruption can be achieved by both mechanical and non-mechanical methods. The disruption process is often quantified by monitoring changes in absorbance, particle size, total protein concentration, or the activity of a specific intracellular enzyme released into the disrupted suspension. Some target products are intracellular, including many enzymes and recombinant proteins. Therefore, potential methods are required to disrupt the microorganisms and release these products.

Breaching of the cell wall/envelope and cytoplasmic membranes can pose problems, particularly where cell possesses strong cell walls. As microorganisms are protected by extremely tough cell walls, so in order to release their cellular components a number of methods for cell disintegration have been developed. For example, disruption of yeast cell preparations typically requires through the pressure cells at 650 bar (1 bar = 100 kPa = 0.987 atm), whereas wild-type *Escherichia coli* generally needs 1100-1500 bar. Although a number of techniques appears to be satisfactory at the laboratory scale but their potentiality remains a constraint at industrial level, particularly in the case of intracellular product isolation.

Gram positive bacteria such as *Lactobacillus* and *Streptococcus* have the simplest cell wall structure consisting of two layers. The outer layer is the murein layer, about 10-80 nm thick, made of peptidoglycan which exist in one form or another in almost all species. The space below the murein layer called periplasmic space, is about 8 nm thick and often contains enzymes. The inner layer called the plasma or cytoplasmic membrane is a double layer made up of phospholipids and some of proteins and metal ions. The plasma membrane controls the permeability of nutrients into the cell and metabolites from the cell into the surrounding solution.

The basic cell envelope of Gram negative bacteria such as *E.coli* and *Pseudomonas sp.* consists of three layers the outer membrane is made of a polymer containing both proteins and lipopolysaccharide. The second layer is

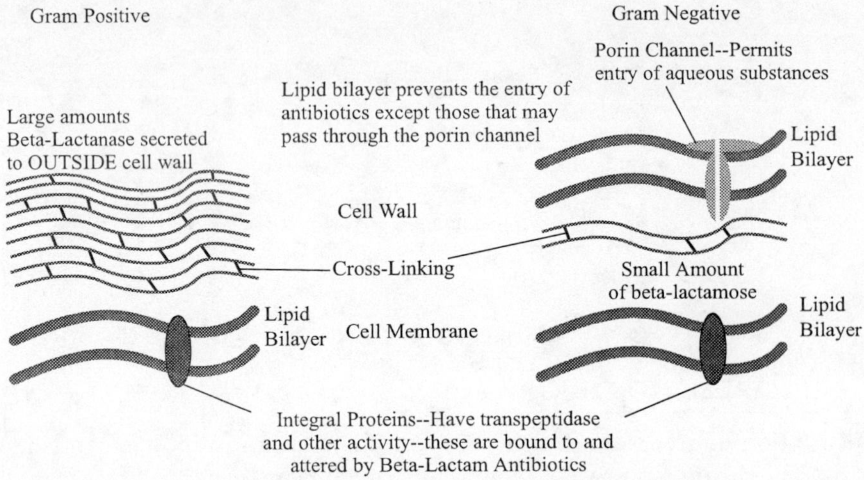

Fig. 2 Gram positive and Gram negative bacteria cell wall

called the murein layer. These two outer layers provide mechanical strength and is necessary to rupture these membranes to release the intracellular product. The third layer is the cytoplasmic membrane, similar in composition and function to the cytoplasmic membrane in Gram negative bacteria.

In prokaryotes, proteins are in solution whereas in genetically engineered plants, excess protein is synthesized and precipated within the cytoplasm. It is necessary to rupture the cell to release the protein in the cytoplasm.

Eukaryotic cells have nuclei and the cell wall is structurally complicated. The cell wall of yeast is thick and is made up of two main layers. The outer layer is a highly cross linked structural layer made up of mannan and protein and below it lies the crosslinked glucan layer. The cytoplasmic membrane beneath the cell wall is a double layer made up of phospholipids. Plant cell walls are mostly made up of cellulose and other polysaccharides. Animal cells do not have a cell wall, but the cytoplasmic membrane is stiffened by sterols.

2.4 CELL DISRUPTION METHODS

Any potential method of cell disruption must ensure that the labile materials are not denatured by the process or hydrolyzed by the enzymes present in the cell. The method that one uses depends largely upon the nature of the cell wall/membrane being disrupted and also upon the nature of the desired product. Various methods available for cell disruption may be classified broadly as: (1) mechanical methods, (2) physical methods, (3) chemical methods (4) enzymatic methods, and (5) combination of two any two or more methods.

2.4.1 Mechanical Methods

Mechanical methods find use in the laboratory as well as in industrial scale operations. In this process, intracellular products are released from microorganisms mainly by mechanical disruption of the cells; the cell envelope is physically broken, releasing all intracellular components into the surrounding medium. Therefore, the product of interest must be separated from a complex mixture of proteins, nucleic acids, and cell wall fragments. The various mechanical methods used are described as follows.

2.4.1.1 Bead mill operation

Either a horizontal or a vertical container is used for stirring cell suspension with glass beads in a completely closed system. Mills come in different designs and sizes. Figure 3 illustrates the various forces involved in the bead mill operation(s).

2.4.1.1.1 Dyano mill

A typical dyano mill used for disruption of cells is composed of a steel chamber with specially designed agitator disks mounted on the shaft which transfers kinetic energy to the spherical beads. An external pump feeds the cell suspension into the mill. Beads can be of glass, zirconium oxide, zirconium silicate, titanium carbide. Beads are retained in the mill by means of a separator/ slot screen. Shape and the arrangement of disks affect the energy input and distribution of beads.

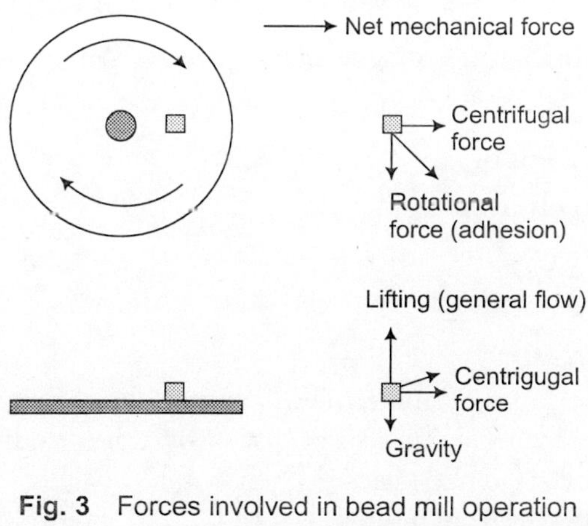

Fig. 3 Forces involved in bead mill operation

The differential velocity will depend on the size and density of the beads. Cylinder is filled 80-90% beads. Within the grinding chamber is a shaft with specially designed discs. When rotated at high speed, high shearing and impact forces from millions of beads break cell walls. The broken cells are then discharged. A primary advantage of bead over pressure devices is better temperature control. During this process heat is liberated, energy dissipation is obtained by spiral flow of cooling water. Also this device is fully enclosed, preventing release of organisms in an aerosol, which is a critical consideration with pathogenic or genetically modified organisms.

2.4.1.2.1 Ball Mill

For a solid medium, such as frozen cell paste or cells attached to or within a solid matrix, different processing options are available. Ball Mills can be operated on large scale processes. More specialized pieces of equipments include the Hughes and X-presses. The Hughes presses consists of a split block with a half cylinder hollowed in each face. The frozen cell paste is placed in the hollow cylinder, and pressure is applied to it from the cylinder into the channels cut in the block. In the X-press, frozen cells are forced continuously through a small hole in the disc between two cylinders at low temperature and pressure. Cell disruption takes place because of deformation of organisms embedded in the ice. Pressure levels of several hundred bars can be obtained with the X-press.

The disruption chamber consists of a conical rotor and conical stator. The gap between the rotor and the stator is partially filled with beads. Cell suspension is pumped in and due to the rotation of the rotor; the cell suspension is forced radially outward. Cooling is provided in the mill stator and lid.

In a bead milling operation the residence time (time interval that the cells are exposed to shear force inside the mill) is important parameter to optimize the protein recovery. It follows a first order kinetics:

$$R = Rm\ (1 - e^{-kt})$$

Taking log and rearranging the equation can also be written as

$$\ln (Rm/Rm - R) = kt$$

Rm = Maximal amount of protein available for release,

R = Protein amount released at certain time,

k = First order constant,

t = Residence time.

2.4.2.1 Homogenizer

The principle of operation is to pump the cell suspension through a horizontal grinding chamber filled with about 80% beads. Within the grinding chamber is a shaft with specially designed discs. When rotated at high speeds, high shearing and impact forces from millions of beads break cell walls. The broken cells are then discharged. A primary advantage of bead over pressure devices is better temperature control. Also this device is fully enclosed, preventing release of organisms in an aerosol, which is a critical consideration with pathogenic or genetically modified organisms.

Disruption at constant pressure

$$R = Rm\,(1 - e^{-kN})$$

Rm = Maximal amount of protein available for release,
R = Protein amount released at certain time,
K = First order constant,
N = Number of passages.

Disruption follows a first order process at a given pressure. The extent of protein released

$$\ln(Rm/Rm - R) = kNP^a$$

Rm = Maximal amount of protein available for release,
R = Protein amount released at certain time,
K = First order constant,
N = Number of passages,
P = Operating pressure,
a = Exponent of pressure.

Homogenizer is often preferred for non-filamentous organisms.

Optimizing the process

1. Increase the operating pressure
   The temperature increase (2.5°C /10MPa), that necessitates cooling.
2. Alter the valve geometry. Instead of flat valve seat, knife edge is preferred.

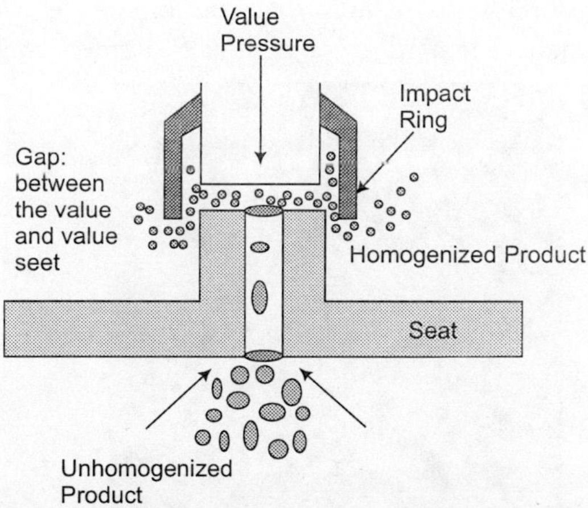

Fig. 4 Schematic diagram of homogenizer used for disruption of cells

APV-Manton-Gaulin Homogenizer

Specific rate of release of enzyme: Periplasmic protein faster than that of the cytoplasmic protein faster than the membrane bound proteins. This is a high pressure positive displacement pump, incorporates an adjustable valve with a restricted orifice. The smallest model has one plunger, while there are several in larger models. During use the microbial slurry passes through a no return valve and impinges against the operative value set at the selected operating pressure. The cells are then passed through a narrow channel between the valve and an impact ring followed by a sudden pressure drop at exit to the narrow orifice. The largest pressure drop across the valve is believed to cause cavitations in the slurry and the shock waves so produced disrupt the cells. The size of the pressure drop is an important factor that helps in achieving effective disruption, and as with all mechanical methods, cell size and shape influence ease of disruption.

2.4.1.3 Agitation with abrasives

Mechanical cell disruption can also be achieved in a disintegrator containing a series of rotating discs and a charge of small beads. The beads are made up of mechanically resistant materials such as glass, alumina ceramics and some titanium compounds. In industries high speed bead mills and disintegrators are equipped with cooling jackets, are often used to agitate a cell suspension with beads. Cell breakage results from shear forces grinding between beads and collision with beads. The efficiency of cell breakage is a function of agitation speed, concentration of beads, beads density and diameter, broth density, flow rate and temperature.

2.4.1.4 Freezing-thawing

Both microbial pastes and plant and animal tissue can be frozen in liquid nitrogen and then ground with a common mortar and pestle at the same low temperature. Presumably the hard frozen cells are fractured under the mortar

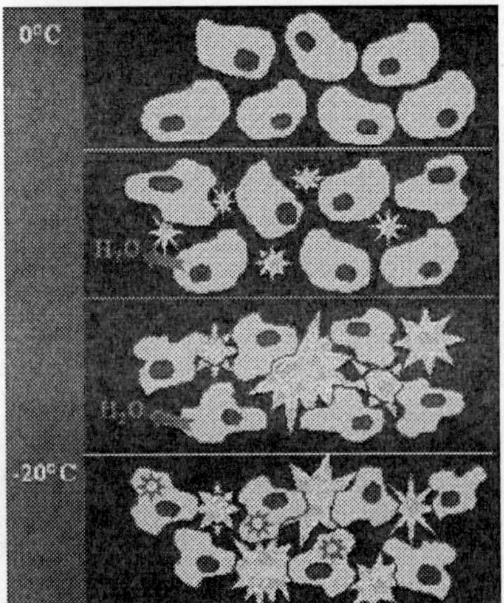

Fig. 5 Process layout for Freezing-Thawing

because of their brittle nature. Also, ice crystals at these low temperatures may act as an abrasive. Freezing and Thawing of a microbial cell paste will inevitably cause ice crystals to form and their expansion followed by thawing will lead to some subsequent disruption in cells (Fig. 5). It is slow with limited release of cellular material and has not often been used as a technique on its own, although it is often been used in combination with other techniques.

2.4.1.5 Ultrasonication

Ultrasonic disruption of cells involves cavitation, microscopic bubbles or cavities generated by pressure waves. It is performed by ultrasonic vibrators that produce a high frequency sound with a wave density of approximately 20 kilohertz/s. A transducer converts the waves into mechanical oscillations via a titanium probe immersed in the concentrated cell suspension. However this technique also generates heat which can denature thermolabile protein and other cell components.

This method can be effective on small scale, but a number of serious drawbacks make it unsuitable for large scale operations. Power requirements are high, there is a large heating effect, so cooling is also needed, and the probes have a short working life and are only effective at a short range. Continuous laboratory sonicators with hold up volume of around 10 cm^3 have been shown to be effective.

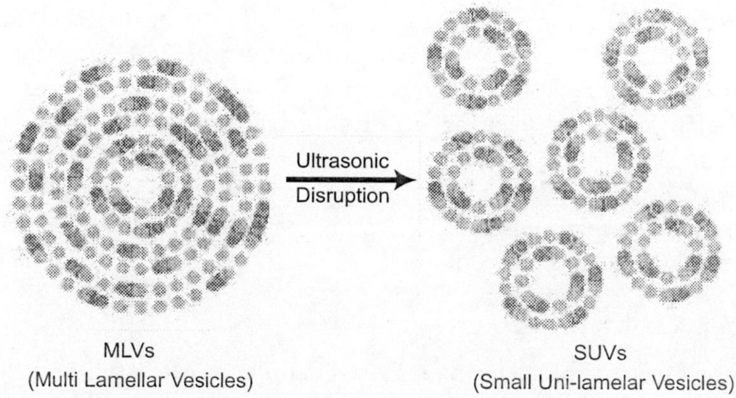

Fig. 6 Ultrasonication of yeast cells

2.4.1.6 Microfluidizer

Cell suspension is driven at a flow rate (250-600 ml/min) by a pump at a constant pressure (20-159 MPa) into an interaction chamber where the stream is split into two, which passes through the microchannels (100 μm) and later directed head-on at each other before emerging at atmosphere pressure. It can be operated in batch or continuous mode.

2.4.2 Physical Methods

In physical method(s) basically large quantities of gases are first dissolved in the cell under high pressure within a suitable pressure vessel. Then, when the gas pressure is suddenly released, the gas comes out of the solution as expanding bubbles that stretch the membranes of each cell until they rupture and release the contents of the cell. The various mechanical methods used are described as follows.

2.4.2.1 Nebulizer

When a gas is blown over a surface of liquid the gas flow causes droplets to be formed, they are connected to the liquid for a millisecond forming a microcapillary. When a liquid is in a capillary tube, flow takes place in layers/ laminar, i.e., in the center it flows faster than in the sides of the capillary. During this flow when a cell encounters this differential speed, it will stretch and break.

Advantages

(a) No cooling is required.

(b) Shear force can be regulated either by gas pressure or by viscosity of the liquid.

2.4.2.2 Decompression

When pressurized microbial cells are gradually penetrated and filled with gas. After saturation, the pressure is suddenly released when the absorbed gas expands leading to rupture. In this method mostly nitrogen gas is used. The nitrogen decompression method is particularly well suited for treating mammalian and other membrane bound cells. It has also been used successfully for treating plant cells, for releasing virus from fertilized eggs and for treating fragile bacteria. It is not recommended for untreated bacterial cells, but this restriction can be eliminated by using various pretreatment procedures to weaken the cell wall. Yeast, fungus, spores and other materials with tough walls do not respond well to this method.

2.4.3 Chemical Methods

Many chemical methods have been employed in order to extract intra cellular components from micro organisms by permeabilizing the outer-wall barriers. It can be achieved with organic solvents that act by creation of canals through the cell membrane: toluene, ether, phenylethyl alcohol DMSO, benzene, methanol, chloroform. Chemical permeabilization can also be achieved with antibiotics, thionins, surfactants (Triton, Brij, Duponal) chaotropic agents, and chelates. The various mechanical methods used are described as follows.

2.4.3.1 Detergents

A wide range of other techniques have been developed for small scale disruption using chemicals and detergents. However some of them can further lead to problems with subsequent purification steps. Basically, the process depends upon pH and temperature. Detergents are amphiphatic, that is they are capable of interacting both with water and lipid. The key mechanism involves the action of detergents in solubilizing the lipids in the cell wall to form micelle. In dilute solutions, the detergents do not dissolve but at higher concentrations lipid solubilization begins suddenly and thereafter increases linearly with detergent concentration. A number of detergents will damage the lipoproteins of the microbial cell membrane and lead to release of intracellular compounds.

The compounds which can be used for this purpose include quaternary ammonium compounds, sodium lauryl sulphate, anionic detergents such as sodium dodecyl sulphate (SDS), sodium sulphonate, sodium taurocholate, cationic detergents such as cetyltrimethyl ammonium bromide (CTAB) and non-ionic detergents such as Triton X-100. But these detergents may cause some protein denaturation and need to be removed before further purification stages can be undertaken. The use of Triton X-100 in combination with guanidine-HCl is widely and effectively used for the release of cellular protein.

2.4.2.2 Organic solvent

Organic solvents usually toluene, benzene, chlorobenzene, xylenes, acetone, butanol, chloroform or methanol, have been used to liberate enzymes and other substances from microorganism by creating channels through the cell membrane. Organic solvents at lower concentrations (1-3%) permeabilize the cell wall without disrupting it. This method is useful in retaining the contents of the cell for sequential release of desired products and also using the permeabilized cells as a porous bag of catalysts. Toluene permeabilization of the cell wall has been seen in the case of *Agrobacterium radiobacter* which produces two enzymes hydantonin and N-carbamylamido hydrolase which are required for the manufacture of semisynthetic penicillins. Ethyl acetate has been used to permeabilize yeast cells to release periplasmic invertase and á-glucosidase.

2.4.3.3 Osmotic shock

Osmotic shock caused by a sudden change in salt concentration will cause disruption of a number of cell types. Osmotic shock to the cells is provided by dumping a given volume of cells into pure water of about twice the volume of cells. The cells swell due to osmotic flow of water ultimately bursting, thereby releasing the products into the surrounding medium. The susceptibility of the cells to undergo disruption by osmotic shock depends on their type. Red blood cells are easily disrupted. Animal cells can be disrupted only after mincing or homogenizing the tissues. Plant cells are more resistant to disruption by this method. Osmotic shock is often minimal on bacterial cells. Osmotic shock is often useful for releasing products from the periplasmic space in gram negative cells. This may be achieved equilibrating the cells in 20% buffered sucrose and then rapidly harvesting and resuspending in water at about 4 degree celcius. The method has been used for releasing hydrolytic enzymes and membrane binding proteins from a number of Gram negative bacteria including *E.coli* and *Salmonella typhimurium*.

2.4.3.4 Alkali treatment

This is a cheap and effective method but very harsh. Alkali acts on the cell wall in a number of ways including saphonication of the lipids. This method is used for the hydrolysis of microbial cell wall material provided that the enzyme will tolerate a pH of 11.5 to 12.5 for 20 to 30 minutes. Proteases are inactivated by this treatment and hence it is possible that the method would be useful in preparation of pyrogen free therapeutic enzymes. This technique has been used for the extraction of L-asparaginase.

2.4.3.5 Antibiotic treatment

Actively growing cells can be treated with antibiotics such as penicillin or cycloserine, which interferes with cell wall synthesis and coupled with the correct osmotic conditions, can lead to cell disruption. Other permeabilization techniques include the use of basic proteins such as protamine: the cationic polysaccharide chitisan is effective for yeast cells: and streptolysin permeabilizes mammalian cells.

2.4.4 Enzymatic Methods

There are a number of enzymes which hydrolyze specific bonds in cell walls of a limited number of microorganisms. Examples of such enzymes are glucanase, mananase, lysozyme etc. Digestion of the cell wall is achieved by the addition of these lytic enzymes to a cell suspension. Enzymes are considered to be highly selective, gentle and most effective, but are costly. Only enzyme, lysozyme has found commercial application in the industry, mainly for the extraction of enzymes particularly glucose isomerase from *Streptomyces sp*.

The enzyme lysozyme, isolated from hen egg whites, hydrolyses á-1,4-glycosidic bonds in the mucopeptide cell wall. The peptidoglycan cell wall can therefore be removed from Gram positive bacteria by treatment with lysozyme, and if carried with suitable buffer, once the cell has been digested the cell membrane will rupture due to the osmotic effect of the suspending buffer.

Gram negative bacteria can similarly be disrupted by lysozyme but treatment with the aid of a detergent such as Triton X-100, or addition of EDTA. Addition of EDTA removes calcium ions, thus destabilizing the outer lipopolysaccharide layer. This effectively permeabilises the outer membrane, allowing the access of the lysozyme to the peptidoglycan layer. If carried in an isotonic medium so that the cell membrane is not ruptured, it is possible to selectively release proteins from the periplasmic space. It is possible to form *E.coli* spheroplasts, cells with intact inner membrane, but porous or completely removed outer membrane by combined use of lysozyme and EDTA.

In addition to lysozyme, three other types of bacteriolytic enzymes have been isolated. These include (1) glycosidases, which split the polysaccharide chains, (2) acetylmuramyl-L-alanineamidases, which cleave the junction between the polysaccharide and peptide, and (3) endopeptidases, which split the polypeptide chain. Some of the proteases also have been shown to be bacteriolytic.

Yeast can be similarly disrupted using enzymes to degrade the cell wall and either osmotic shock or mild physical force to disrupt the cell membrane. Enzymatic digestion alone allows the selective release of proteins from the periplasmic space. A mixture of two types of enzymes, a lytic protease which lyses the outer mannoprotein layer of the cell wall exposing the inner glucan layer which is acted on by the second enzyme glucanase is useful to obtain a sufficiently large hole in the cell wall, facilitating the extrusion of the plasma membrane and its contents intact as protoplast in osmotic support buffers containing sucrose or mannitol. In dilute buffers, the protoplast also lyses releasing the cytoplasmic proteins and organelles which may also lyse. Plants cells can be lysed by cellulose and pectinase.

Mechanical methods are not specific and tend to release the desired product along with a host of other biomolecules and cell debris. Enzyme lysis of microbial cells has advantages in that controlled sequential degradation of the cell wall can give rise to selective product release their by preventing their contamination and denaturation.

Sequential disruption of microbial cells for selective product release involves the use of lytic enzymes depending on parameters such as the type of microbial cells, product to be harvested and process condition.

3

CHAPTER

FILTRATION

3.1 INTRODUCTION

Filtration is the process to mechanically separate a mixture of solids and fluids using a filter medium or filter cloth which retains solid or it is also defined as a pressure driven separation process that uses membranes to separate components in a liquid solution or suspension based on their size and charge differences. Depending on the application, the solid, the fluid, or both may be isolated. Either the clarified fluid or solid removed form fluid may be the desired product.

The art of filtration was known to primitive man, who obtained clear water from a muddy river by scooping a hole in the sand in a river bank to a depth below the river water level. This process was carried out for 100 years to purify water.

The basic requirements for filtration are:

- A filter medium
- A fluid with suspended solids
- A driving force such as a pressure difference to cause fluid to flow
- A mechanical device called filter that holds the filter medium, contains the fluid and permits the application of force. Filter may have special provisions for removal of the filter cake or the solid particle

Filter medium allows the fluid to pass through and retains the solid. The separated solid called the filter cake forms a bed of particular filter medium. The thickness of cake increases from zero to final thickness at the end of filtration. During filtration therefore fluid first passes though filter medium and then with filter cake.

3.2 PRINCIPLE

Filtration, method to separate solid and liquid is done by applying driving force. Fluid to be filtered is allowed to pass through filter medium (often known as filter bed), only if some driving force is applied. This force may be caused by gravity, centrifugation, application of pressure on the fluid above the filter or application of vacuum below the filter or by a combination of such forces. Gravitational force alone may be used in large sand-bed filters and in simpler laboratory filtrations, which are commonly termed as barrier filtration (Figure 1). Centrifuges containing a bowl with a porous filter medium may be considered as filters in which gravitational force is replaced by centrifugal force. If a laboratory filtration is difficult a partial vacuum is usually applied to the container below the filter medium to increase the rate of filtration. Most of industrial filtration process involves use of pressure or vacuum.

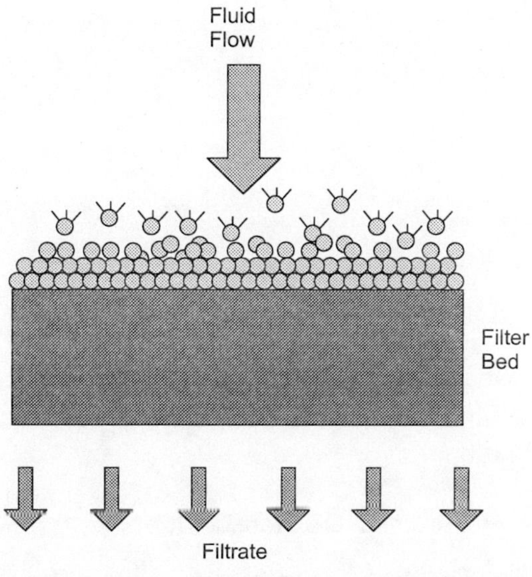

Fig. 1 Diagram showing barrier filtration or simply filtration

Ease of filtration use depends on the properties of solid and fluid. Filtration of crystalline, incompressible solids in low viscosity liquids is relatively straightforward. In contrast, fermentation broths can be difficult to filter because of the small size and gelatinous nature of the cell and the viscous behavior of the broth.

3.3 FILTRATION THEORY

First step in filtration is to apply the full pressure which can be applied by a filter and pressure should be made constant throughout the run. If the initial pressure is high, the first particles caught in the cloth may be compacted into tight mass of low permeability which results in low rate of filtration. If the sludge is not homogenous and contains both crystalline and colloidal particles, the colloidal portion may be forced into the voids between the crystalline solids, greatly decreasing rate of filtration.

On the other hand, a low initial pressure may mean that the first filtrate will not be clear as the cake is formed with a more open structure. Therefore, the common method of filtration is to filter at a constant rate of flow of filtrate during the early part of the cycle. As soon as cloths are well coated with cake and the filtrate is clear, the pressure is increased to the maximum and the filtration is continued and completed under conditions of constant pressure.

At any instant during filtration, rate of filtration is given by the equation:

$$\frac{1}{A}\frac{dV_f}{dt} = \frac{p}{\mu_f[\mu(M_c/a) + r_m]} \qquad \text{...(3.1)}$$

where;

A = Filter area,

V_f = Volume of filtrate,

t = Filtration time,

p = Pressure drop,

μ_f = Filtrate viscosity,

μ = Liquid viscosity,

M_c = Total mass of solids in the cake,

a = Average specific cake resistance,

r_m = Effect of the filter cloth and any particles wedged in it during initial stages of filtration,

$\frac{dV_f}{dt}$ = Filtrate flow rate or volumetric rate of filtration

It is useful to improving rate of filtration for a given batch of material. Various strategies can be deduced from the relationship between variables in equation (1) are:

1. *Increase the filter area (A):* When all other parameters are constant, if A is increased, rate of filtration increases. However this need larger equipments thus increase cost of filtration
2. *Increase the filtration pressure drop (p):* The problem is that 'a' increases with 'p' as indicated in for compressible cakes. Thus improvement can only be achieved from pressure drop by reducing compressibility of cake.
3. *Reduce the cake mass (M_c):* This is achieved by reducing the thickness cake deposited. This is mainly done in continuous rotary filtration by reducing cake deposit per revolution

4. *Reduce the liquid viscosity* (μ): Materials are diluted to reduce viscosity.
5. *Reduce the specific cake resistance* (a): Method to reduce specific cake resistance are:
   (a) Increase the porosity: Cake porosity usually decreases as cells are filtered. Application of filter aid reduces this effect.
   (b) Reduce the specific surface area of the particle: Change in fermentation condition and broth pretreatment are used to achieve these effects.

3.4 TYPES OF FILTER

Filter may be classed according to the nature of the driving force that causes filtration (i.e. gravity filters, pressure filters and vacuum filters). They also are described according to mechanical characteristics (i.e. plate and frame filters, rotary drum filters). Filters may operate on either batch or continuous basis.

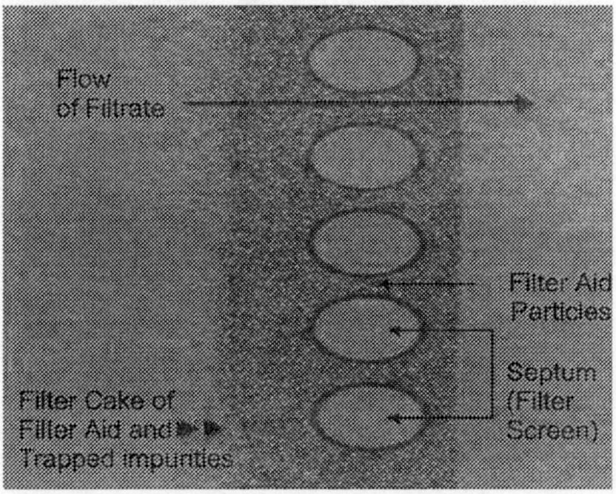

Fig. 2 Filtration of slurry using filter aids

3.4.1 Gravity Filter

The gravity filter is the oldest and simplest one. Gravity sand bed filters installed in the city water plants filter more gallon of water than any other type of filter. Such filters involve the use of tanks that usually are constructed of concrete. At the bottom of the tank is a grating or false bottom; above this is coarse gravel or crushed rock of graded size; at the very top are layers of quartz sand of uniform size.

They are widely used to process large quantities of fluid containing small quantities of solids in water purification. The tanks may be constructed of wood, steel, or other suitable metal but for water purification concrete is used.

3.4.2 Pressure or Vacuum Filters

Pressure or vacuum filters usually are used in industry in preference to gravity filters. The driving force can be supplied by pressure or vacuum is much greater than gravity, thus permitting higher filtration rates. Sand-bed filters are operated under pressure in closed vessels. Pressure or vacuum filters with a thin filter medium such as filter cloth or woven metal screen are preferred to filter slurry containing high solid content.

Advantages

- Cakes are obtained with very low moisture content
- Cakes may be disposed and flattened in layers provided they are not thixotropic and sufficiently stable to support a tractor.
- Intensive soluble recovery and removal of contaminants from the cake may be achieved.
- Solutions may be polished for a higher degree of clarity
- The filter bodies and internals may be constructed from a wide variety of alloys including synthetic materials to filterpress
- Pressure filters are widely available from labor intensive operation to fully automatic machines

Disadvantages

- Cloth washing is difficult and if the solids are sticky a 3mm precoat from diatomite or perlite is required. This adds another step prior to filtration and if not done carefully and too thin precoat is formed it may leave areas with exposed cloth which will bind rapidly. Conversely, if a too thick precoat is used, then it will consume effective cake volume.
- The operator can hardly see the forming cake and is unable to carry out an inspection while the filter is in operation.
- The internals are difficult to clean and this may be a problem with food grade.

The most common type of pressure filter with a filter cloth is known as filter press. This is a batch-operated filter that is used when the filter capacities involved do not warrant investment in more expensive continuous or vacuum filters.

3.4.3 Plate and Frame Filters

Plate and Frame Filters is a pressure filter in which the simplest form consists of plates and frames arranged alternatively. The plates are covered with filter cloths or filter pads. The plates and frames are arranged on horizontal framework and held together by means of screws or hydraulic ram so that there is no leakage between plates and frames which forms liquid tight compartments.

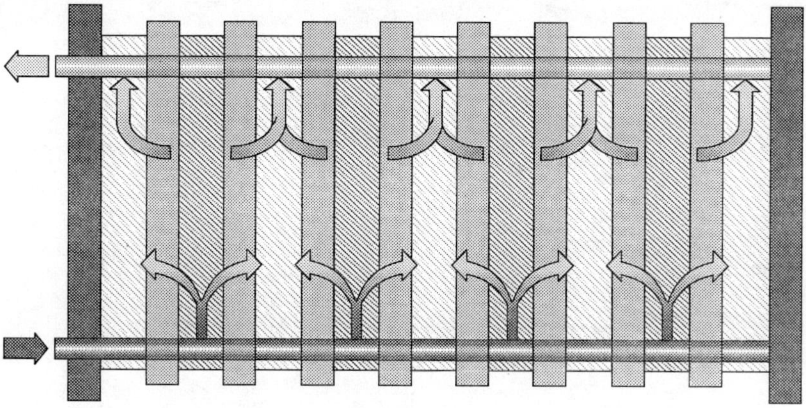

Fig. 3 Plate and frame filters are used for filtration

The slurry is piped under pressure to the filter frame usually located at the corners of the plates, filtrate passes through the filter cloth or pad and runs down to the grooves in the filter plates and is then discharged through recessed channels and additional ports in the plates. The solids remained in the frame and filtration stops when the frames are completely filled. After a filter cake has been removed deposited on the cloth it may be washed by pumping water through the filter, then the plates are separated and the filter cake is removed manually. Plate and frame filters press requires the least floor space per unit of area. However, since it is batch operated, it must be loaded and unloaded manually in the labor costs are high, particularly in the larger sizes.

3.4.4 Pressure Leaf Filters

Pressure Leaf Filters are also used for pressure filtration on a batch basis. This filter incorporates number of leaves, consisting of a centre section of coarse metal mesh that supports the filter medium and permits the filtrate to escape. The centre section is covered on either side with the filter medium which usually is a woven wire screen of appropriate mesh size. Leaves may be circular or rectangular and are assembled on frame with ports for removal of the filtrate. The assembly is usually enclosed in a pressure vessel that also contains the slurry after sufficient filter cake is deposited on the leaves, the pressure removed, the vessel opened and the leaf assembly with filter cake is taken out. The cake is removed by water spray or by applying air pressure to filtrate line. The leaf filter as compared to filter press, costs less to operate because of the advantageous provision for efficient washing and removal of cake without removing the leaves.

Vertical metal-leaf filter: This filter consists of a number of vertical porous metal leaves mounted on a hollow shaft in a cylindrical pressure vessel. Filtrate

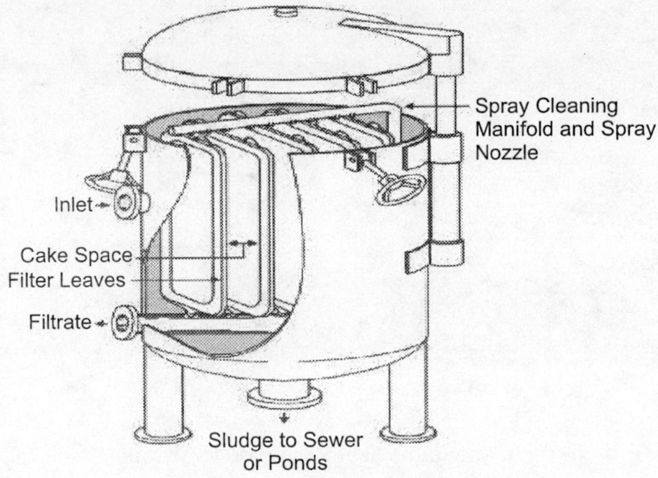

Fig. 4 Vertical metal leaf filter

is removed from plates via the horizontal hollow shaft. In some designs hollow shaft can some time be slowly rotated during filtration. Solids are normally removed at the end of a cycle by blowing air through shaft and into the filter leaves.

Horizontal metal-leaf filter: In this, metal leaves are mounted on a vertical hollow shaft with in a pressure vessel. Often, only the upper surface of the leaves is porous. Filtration stops when cake fills the space between leaves. At the end of a process, the solid cake can be discharged by releasing the pressure and spinning the shaft with a drive motor. Figure 5 depicting horizontal pressure filter is a type of horizontal metal-leaf filter.

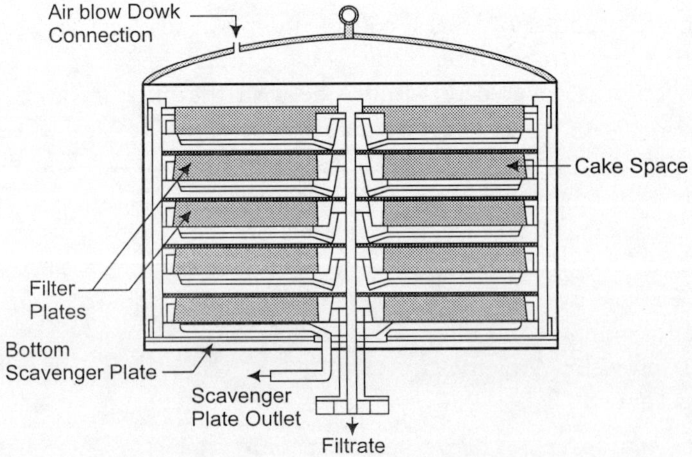

Fig. 5 Horizontal pressure filter

3.4.5 Stacked-disc Filter

In Stacked-disc filter there is no filter cloth and the bed is easily replaced, and undoubtedly labor costs are lowered significantly. It consists of number of precision-made rings which are stacked on a fluted rod. The rings are made of stainless steel and are assembled on the rods. Diatomite is generally used as filter aid. During filtration, filtrate passes through disc and is removed through the grooves of fluted rods, while solids deposited on the filter coating and is removed from the rings by applying back pressure via the fluted rods. Horizontal pressure filter shown above in figure 5 is a type of stacked-disc filter.

3.4.6 Rotary Vacuum Filters

Rotary vacuum filters are used extensively in industry for continuous filtration of large quantities of slurries containing high content of solids. The filter consists of rotating, hollow, segmented and cylindrical drum with internal divisions, ports and valves for application of the vacuum and removal of filtrate. The drum is covered with the filter medium, usually a woven wire screen or cloth and is partially submerged in the trough of slurry. The partial vacuum inside the drum causes the filtrate to flow into the drum and out of the ports, depositing a filter cake on the surface of the drum. As the drum rotates, the cake may be sprayed with wash water and then removed by scrapers.

A number of rotary vacuum filter are manufactured, which differs in mechanism of cake discharge from the drum.

1. *String discharge:* The cake is lifted free from upper part of drum when the vacuum pressure is released and carried to the smaller rollers where it falls free. Mainly fibrous filter cake is separated from string discharge.
2. *Scraper discharge:* The filter cake which builds up on the drum is removed by an accurately positioned knife blade, because knife is close to the drum, as there occurs the gradual wearing of filter cloth on the drum. In scraper discharge with pre-coating on drum, the filter cloth on the drum can be blocked by bacterial cells. This problem is overcome by precoating the drum with a layer of filter aids 2 to 10 cm thick. The cake which builds up on the drum during operation is cut away by the knife blade which slow mechanically advances towards the drum towards at a controlled rate. Alternatively, the blade may be operated manually when there is an indication of 'blinding' which leads to reduction in filtration rate. In either case cake is removed together with a very thin layer of precoat for better rate of filtration.

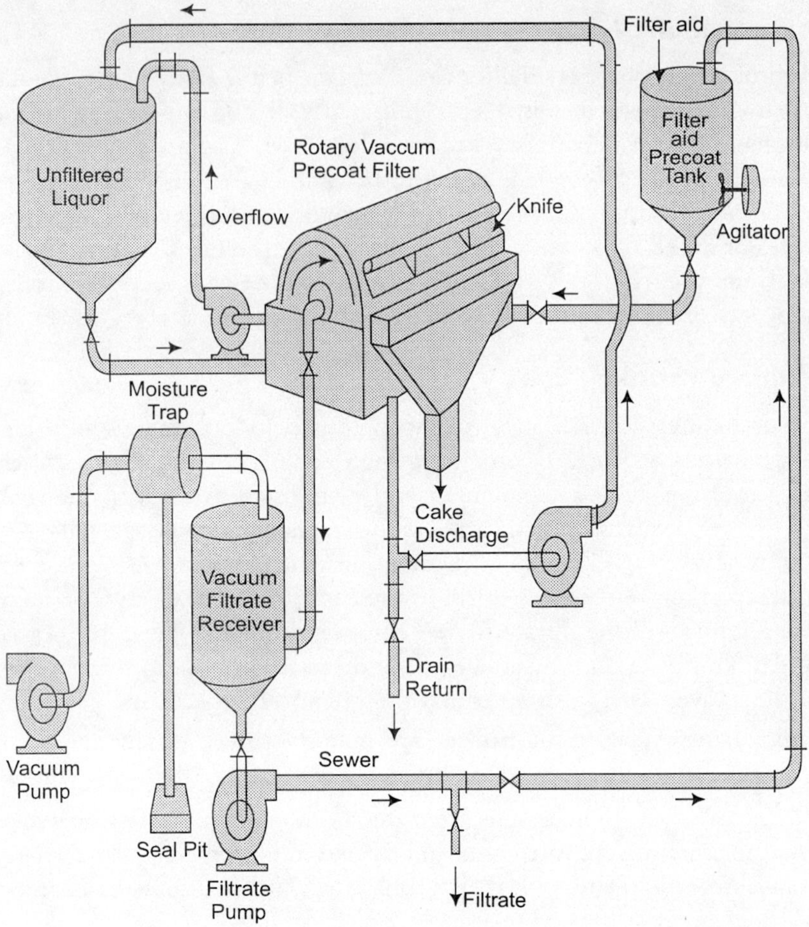

Fig. 6 Rotary vacuum filter

3.5 FILTER AIDS

One of the most widely used and robust means of microparticulate separation in liquid system is highly porous powdered media, often called filter aids. Filter aids are fine, chemically inert powders applied in both process and waste microfiltrations to maintain high flow rates while giving brilliant clarity. Powdered media provides versatility, high solids loading capacity, high throughout or high clarity, ease of scale-up and frequently the lowest cost option in any filtration process. Filter media offers high porosity, only about 15%of the total volume is solid rest is empty space. Such high porosity facilitates liquid flow around the particles and improves rate of filtration.

Filter aids can be used as precoat on the filter medium to protect against the penetration of unwanted solids and prevent blockage of filter by solids. Filter aids are also used to increase the porosity of the cake as it forms. The addition of filter aids increases permeability of cake provides channels for filtrate recovery and extends cycle length and restricts solids movement.

Porous Media (Filter Aid) Types: Some most widely used porous media (filter aids) in filtrations are:

Diatomite: Diatomite is obtained from diatomaceous earth, sediment greatly enriched in biogenic silica form of the siliceous frustules of diatoms, adverse array of microscopic, single-celled algae of the class *Bacillariphycae*. Diatomite products are characterized by an inherently intricate and highly porous structure composed primarily of silica, along with impurities of alumina, iron oxide and alkaline earth oxides.

One major advances in diatomite product technology is the development of Celpure, a highly diatomite with enhanced performance. During manufacture of Celpure the impurities normally present in filter aids are removed thereby yielding a high silica content and ultra pure product. Celpure has lighter density. This enables using less filter media (by weight) to match or surpass the performance of an existing filtration process.

Perlite: Perlite is a naturally occurring volcanic glass which thermally expands upon processing. Perlite is chemically a sodium potassium aluminum silicate. After milling, a porous, complicated structure is present, but because its structure is not as intricate as that of diatomite, perlite is better suited for separation of coarse microparticulate from liquids having high solids loading

Demerits of Filter Aids

Besides having great importance in filtration, filter aids have some demerits and these are

- They add to the cost of filtration; minimum quantity needed to achieve desired result must be established experimentally.
- Some filter aids like diatomite absorbs liquid, so when fermentation product is in liquid phase some will be lost.
- Filtration rate increases but filtrate clarity decreases with the implementation of filter aids.
- Disposal waste cell is difficult if it contains diatomite.

3.6 BATCH FILTRATION

In pressure filters which are batch-wise in operations, the whole of the filter medium surface in each unit is going through a cyclic process of cake formation. This leads to discontinuous solid discharge from the filter, usually taking place of during that part of the cycle when the filter is taken off pressure and opens to atmosphere.

Batch filtration consists of cycles, these are:

Filtration: The most important feature of the pressure filters which use hydraulic pressure to drive the process is that they can generate a pressure drop across the medium of more than 1 bar, which is the theoretical limit of vacuum filters. Whilst the use of high pressure drop is advantageous, leading to higher outputs, drier cakes or greater clarity of overflow. This is not necessarily case, in compressible cakes, an increase in pressure drop leads to decrease in filtration rate and also decrease in permeability of cake.

Dewatering: Dewatering is done through the process of passing air through filter cake, but this includes a threshold pressure to be exceeded in order that air may enter the filter cake. There is minimum moisture which can not be removed at any pressure, called irreducible saturation.

Optimization of cycle times: In batch filters, one of the important decisions to be made is how much time is to be allocated to the different operations such as filtration, dewatering, cake washing and cake discharge. All of this will have to be happening within time, which itself is not fixed. If all the non-filtration operations are grouped together into a 'downtime', and this is assumed to be fixed and known, we can derive an optimum filtration time.

Cake squeezing: The compressibility of many industrial filtration cakes, an unwanted phenomenon in filtration can be turned advantageous by mechanical squeezing of cake which acts as additional dewatering method. The fact that a cake has a reasonable or high porosity does not necessarily mean that the liquid with in it can be easily removed. Some of the liquid can, however, be removed simply altering the cake structure by mechanical compression. In fact, the more compressible materials are often also less permeable. It is therefore, the very cakes which consolidate under hydraulic pressure and render such filtration are the best candidates for mechanical squeezing. Compressibility of cakes can be often promoted by addition of flocculating agent. Those pressure filters which use the squeezing of the cake are, on the whole, expensive and their use can be justified only if the energy savings large enough. There are other positive features of squeezing of the cake in that it often improves cake release, it closes any cracks and tightens the cake thus making any cake washing more effective.

3.7 CONTINUOUS FILTRATION

In continuous filtration there is no need to discharge the cake, it itself gets discharged. Continuous filters are meant to be operated at high pressure and high speeds. The first advantage of continuous filters is the absence or reduction in down time which means overall gain in the capacity. In continuous filtration, dry cake production Y is proportional to the square root of the pressure drop p and to the square root of the speed of rotation, n:

$$Y = \text{constant}.\sqrt{Pn}$$

This equation is only valid for incompressible cakes, negligible medium resistance and all others parameters are constant.

The increased pressure drop in pressure filtration is in two ways: It increases capacity directly through higher driving force and it allows the speed of rotation *n*, to be increased without increasing moisture contents (because higher pressure leads to the better dewatering). Higher speeds leads to thinner cakes and the limiting condition is the minimum cake thickness that can be successfully be removed from the filter medium and discharged. Thin cakes offer little resistance to air flow and air rates becomes excessive. It seems therefore that continuous pressure filters should be built to run at high pressure and high speeds. There are however two considerations:

- The higher the design pressure, the more expensive is the pressure vessel with its associated hardware.
- High pressure leads to excessive air rates and energy consumptions.

Therefore there must be compromise between cake thickness and pressure.

A continuous pressure filter can be defined as follows: it is a filter which operates at pressure drops greater than 1 bar and its operation does not need to be interrupted to discharge the cake. The cake discharge itself, however does not necessarily continuous and there is little or no down time involved.

Most of the continuous pressure filters available today have their roots in vacuum-filtration technology. It is obvious that a rotary drum or rotary-disc vacuum filter can be adapted to pressure by enclosing it in a pressure cover. The enclosure is a pressure vessel which is heavy and expensive, the progress of filtration cannot be watched and removal of cake from vessel is most difficult problem of the whole design. Other complications of this method are caused by the necessity of arranging for two or more differential pressures between the inside and outside of the filter. Despite these disadvantages, the advantages of high throughputs and low moisture content in the filtration cakes have justified the recent and vigorous development of continuous pressure filters.

3.8 CROSS FLOW FILTRATION (MEMBRANE FILTRATION)

Cross flow filtration is often termed as cross flow membrane filtration, or tangential filtration. In cross-flow filtration flow of medium to be filtered is tangential to the membrane (Figure 7) and no filter cake builds up on the membrane. Merits which cross-flow filtration carries along with it are:

i. Efficient separation.
ii. Closed system for containment of organisms with no aerosol separation is independent of cells and media densities.
iii. No addition of filter aid.

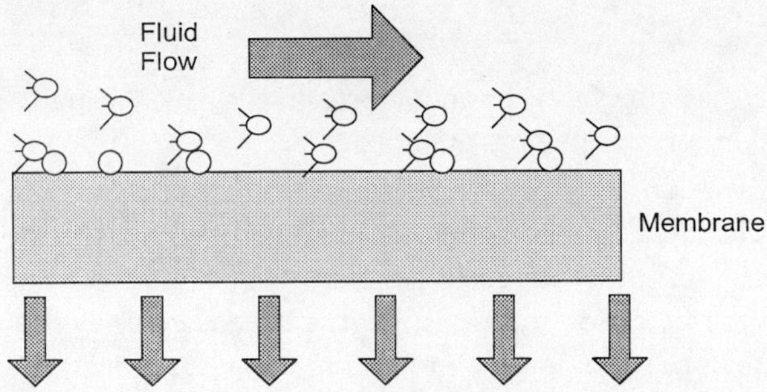

Fig. 7 Generalized diagram of cross flow membrane filtration or tangential filtration

Major components in cross-flow filtration are media storage tank, a pump and a membrane pack. Two types of membrane are used mainly; microporous membranes with a specific pore size and ultrafiltration membrane.

The output from the pump is forced across the membrane surface; most of this flow sweeps the membrane returning retained species back to the storage tank and generally less than 10% of the flow passes through the membrane. As this process is continued, the cells or retained species are concentrated to between 5 to 10% of their initial volume. Various parameters can be modified to optimize flux across the membrane during cross-flow filtration. The first parameter is the transmembrane pressure, which is defined as the difference in pressure between the filtrate side of the membrane and the permeate side of the membrane. The transmembrane pressure is the driving force for the membrane separation.

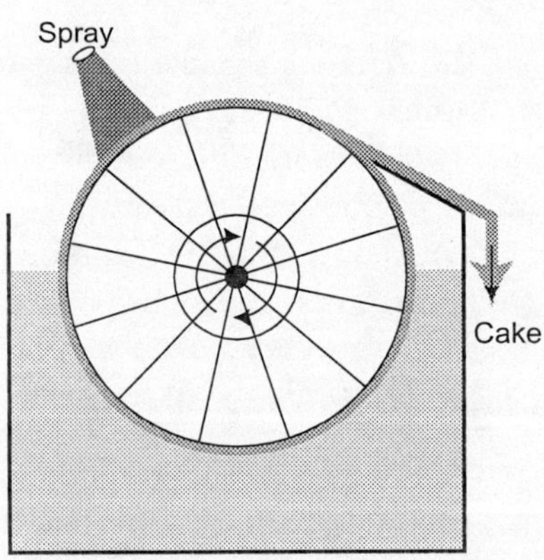

Fig. 8 Cross flow filtration

In general, an increase in the transmembrane pressure increases the flux across the membrane. The exception to this generalization occurs if a compressible filter cake is present on the surface of the membrane. A second parameter which can be varied is the shear rate across the surface of the membrane. Shear rate in cross-flow filtration systems is essentially the velocity at which the filtrate is passed across the surface of the membrane. This shear acts to sweep away particles which may have accumulated at the surface of the membrane (Figure 8). During cross-flow filtration of a non-protein containing solution, increased transmembrane pressure and increased shear generally tend to increase the flux across the membrane. During the filtration of protein solutions, however, increased transmembrane pressure and increased shear usually act to reduce the flux across the membrane. This is due to the structure and nature of the proteins within the solution.

4

CHAPTER

CENTRIFUGATION

4.1 INTRODUCTION

Centrifugation is one of the most important and widely applied research techniques in biochemistry, cellular and molecular biology, and in medicine. Current research and clinical applications rely on isolation of cells, subcellular organelles, and macromolecules, often in high yields. Further, an important task for studying the biochemical and physiological properties of organelles and biomolecules is the preservation of their biological functions and properties during the separation of cellular components. A key technique for separating and analyzing the various elements of a cellular homogenate is represented by centrifugation.

A centrifuge uses centrifugal force (g-force) to isolate suspended particles from their surrounding medium on either a batch or a continuous-flow basis. Applications for centrifugation are many and may include sedimentation of cells and viruses, separation of subcellular organelles, and isolation of macromolecules such as DNA, RNA, proteins, or lipids. Centrifugation is a process used to separate or concentrate materials suspended in a liquid medium, and theoretical basis of this technique is the effect of gravity on particles (including macromolecules) in suspension. For Example, in a solution two particles of different masses will settle in a tube at different rates in response to gravity. Centrifugal force (measured as xg, gravity) is used to stimulate this settling rate in an instrument called a centrifuge. It has been observed that centrifugation of fermentation broth produces highly viscous, concentrated cell sludge, and as a consequence of that most of commercial centrifuges generate large amount of heat due to friction. So requisitely such centrifuges should have a proper amount of ventilation and cooling during their operation.

Two most common examples of the use of centrifugal force are: (1) When you do the "around the world" trick with a yo-yo, it is centrifugal force that makes the yo-yo body stay at the end of the string as you rotate it; and (2) When you wash clothes in a washing machine, it is centrifugal force generated in the "spin" cycle that forces water out of the fabric to facilitate faster drying. Centrifugation may be used to separate materials of different density, remove cells from fermentation broth, to eliminate cell debris, to collect precipitates. Generally centrifugation is found to be effective when undesirable particles, which are required to be seperated, are too small and filtration appears to be a daunting task.

4.2 HISTORY OF CENTRIFUGATION

A **centrifuge** is a piece of equipment, generally driven by a motor that puts an object in rotation around a fixed axis, applying force perpendicular to the axis. The centrifuge works using the sedimentation principle, where the centripetal acceleration is used to separate substances of greater and lesser density.

Fig. 1 Theodor Svedberg

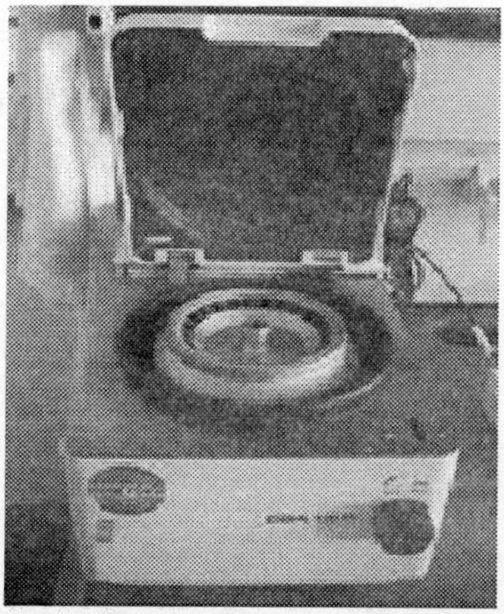

Fig. 2 A laboratory tabletop centrifuge

Centrifuges were first used in the mid-1800s for processes such as separation of milk. These early machines were hand driven and could reach speed up to 3000rpm. Friedrich Miescher, a physiological chemist from Basel Switzerland, was the first person to use a centrifuge to isolate cell organelle. In 1869 he used a similar process to isolate cell nuclei and became attributed with discovering DNA. By 1912 new centrifuges were developed which ran on electrical power.

In 1926 Theodor Svedberg (1884-1971) received a Nobel Prize for his work in colloid chemistry and invention of the ultracentrifuge. During the 1920's-

30's his group built a number of ultracentrifuges capable of reaching 900,000xg. These models had very small rotors. Larger rotors were used for routine work. Svedberg used his centrifuge to determine the molecular weights and subunit structure of highly complex proteins such as haemoglobin and the results of these studies shaped new ideas about the structures of proteins. His later work in nuclear chemistry helped improvements of the cyclotron. Svedberg also aided in the development of electrophoresis by one of his students.

During the early 1930s Martin Behrens worked to improve density gradient techniques involving centrifugation. He has been credited with the introduction of density gradient centrifugation as a means of separating and isolating cells and subcellular components. In 1942, Dr. Albert Claude and Dr. James S. Potter published a paper entitled "Isolation of Chromatin Threads from the Resting Nucleus of Leukemic Cells". In this landmark work they outlined a process involving a series of centrifugation steps and during these steps either the supernatant or the sediment was collected until the finally "chromatin threads" were retrieved from the final sediment.

4.3 PRINCIPLE AND THEORY

In a solution, particles whose density is higher than that of the solvent sink (*sediment*), and particles that are lighter than it *float* to the top. The greater the difference in density, the faster they move. If there is no difference in the density (isopyknic conditions), the particles *hover*. To take advantage of even tiny differences in density to separate various particles in a solution, gravity can be replaced with the much more powerful "centrifugal force" provide by a centrifuge.

Basic principle of centrifugation relies on the effect of *sedimentation* which is due to the influence of gravitational field and centrifugal field. In a centrifuge it is important to know that: (1) Denser a biological structure, the faster it sediments, (2) More the frictional coefficient, the slower particles will move, and (3) Greater the centrifugal force, the faster the particle will sediment. The frictional coefficient depends upon the size and shape of the particle .The frictional force of a particle passing through the viscous fluid is the product of its velocity and its frictional coefficient .The frictional force acts in a direction opposite to sedimentation. Sedimentation of a molecule is influenced by:

- Properties of the molecules (size, shape, density)
- Properties of the solvent, or gradient material (density, viscosity, temperature)
- Interactions between the solute molecules and the solvent gradient material

Centrifuges are devices used in a variety of scientific and technical applications which spin carrier vessels (centrifuge tubes) at high rotation speeds and very high centrifugal force. The centrifugal force (expressed as # *gravities* or, # xg) generated is proportional to the rotation rate of the rotor (in rpm) and the distance between the rotor center and the centrifuge tube. Therefore, a given centrifuge may use multiple rotor sizes to give flexibility in choosing centrifugation conditions. Each centrifuge has a special graph, a nomograph, or a table which relates rotation rate (rpm) to centrifugal force (xg) for each size of rotor it accepts.

Typically, the material to be "spun" is placed in a centrifuge tube which is then placed in a rotor. The rotor is a generally a dense metal which dissipates heat quickly, and is of sufficient mass that it generates momentum, i.e., once its spinning it requires little energy to keep it going. Centrifuges generally work under vacuum and are refrigerated to reduce heating caused by frictional forces as the rotor spins. Rotors are usually stored in refrigeration units to keep them at or near the operating temperature. Because centrifuges come in all shapes and sizes, and the rotors vary, the universal and transferable unit of centrifugation is centrifugal force (xg). In lab write-ups you should report the centrifugation force used (#gravities) because it is the transferable unit between different centrifuges.

As the rotor spins in a centrifuge, centrifugal force is applied to each molecule in the sample:

$$\textbf{Centrifugal force } (F_c) = m\omega^2 r$$

Where, *m* is the mass (particle weight, or molecular weight); ω (omega) is angular velocity (radians/sec) and *r* is the distance from the axis of rotation. This equation says that the larger the molecule, or the faster the centrifugation, or the longer the axis of rotation, the greater the centrifugal force and the rate of sedimentation.

Two forces that act opposite to the centrifugal force are the buoyant force (or displacement force) and the frictional force. The buoyant force represents the force it takes to displace solvent as the particle moves through the centrifugal field. The frictional force represents the drag on the particle as it passes solvent molecules. These two forces are respectively defined as:

$$\textbf{Buoyant force } (F_b) = -m_o\omega^2 r$$

$$\textbf{Frictional force } (F_f) = -fv$$

Where m_o = the mass of the displaced solution, f = frictional coefficient and v = velocity of the particle. When molecule sediments it moves faster and faster in a centrifugal field until the centrifugal force equals the counteracting buoyant and frictional forces:

$$\text{Centrifugal force } (F_c) = \text{buoyant force } (F_b) + \text{frictional force } (F_f)$$

This occurs because the frictional force increases with increased rate of sedimentation whereas centrifugal force and buoyant force are constant for a particular molecule and rotor speed. The particle will move at a velocity such that the total force equals 0, thus:

$$F_c + F_b + F_f = 0, \text{ or}$$

$$m\omega^2 r - m_o\omega^2 r - fv = 0;$$

Substituting $m_o = mv\rho s$, where v is the partial specific volume of the particle and ρs is the density of the solvent, and solving for v results in:

$$v = \omega^2 rm\ (1 - \nu\rho_s)/f, \text{ or}$$

$$v = \omega^2 rm\nu\ (\rho_p - \rho_s)/f$$

This equation (expressed in the two different forms) tells us several things about sedimentation:

1. The more massive a particle, the faster it moves in a centrifugal field.
2. The denser a particle (i.e., the smaller its ν) the faster in moves in a centrifugal field.
3. The denser the solution, the slower the particle will move in a centrifugal field.
4. The greater the frictional coefficient (factors such as viscosity, particle shape, and etc. influence this parameter); the slower the particle will move.
5. The particle velocity is 0 when the solution density is greater than the particle density.
6. The greater the centrifugal force ($\omega^2 r$) the faster the particle sediments.

Table. 1 Terms constituting to centrifugation

g: Gravitational acceleration	$g = \omega^2.r_{eff}$	s: Sedimentation coefficient ($s = 10^{-13}$ s)
v: Sedimentation velocity (cm. s–1)	$v = \omega^2.r_{eff}.s$	M: Molecular mass
		$\bar{v}$: Partial specific particle volume (cm^3. g^{-1})
ω: Angular velocity (rad. s^{-1})	$s = \frac{M\cdot(1-\bar{v}\cdot r)}{f}$	r: Density of the solution (g.cm^3)
r_{eff}: Effective radius (cm)		f: Coefficient of friction

The velocity per unit force will be defined as the sedimentation coefficient (*s*), or:

$$s = v/\omega^2 r = m\ (1 - \nu\pi s)/f$$

When mass is expressed in g and *f* in g/sec *s* ranges from 10^{-13} to 10^{-11} sec. This is normally expressed in **Svedberg** (S) units where 1 S = 10^{-13} sec. The

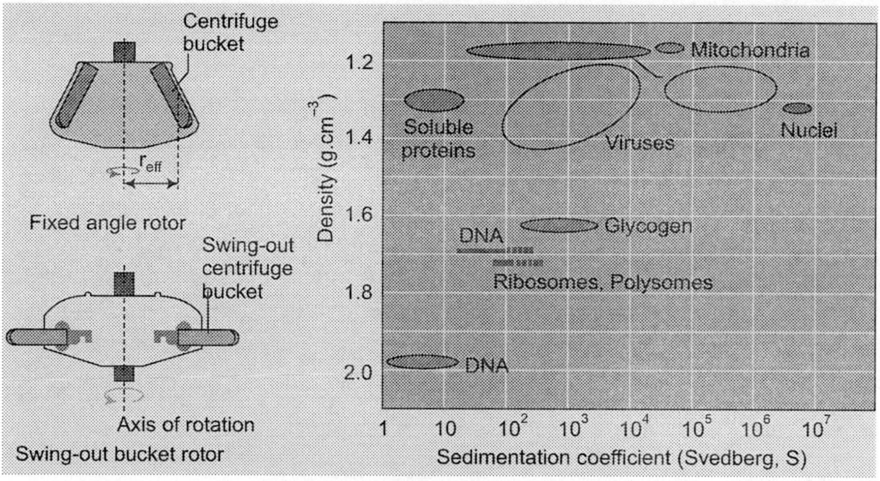

Fig. 3 Principles of centrifugation

higher the *S* value, the faster those particles will sediment. For example, subcellular compartments and macromolecules (Refer to Figure 3) have sedimentation coefficients which reflect their size, shape and density.

The sedimentation coefficient is determined by measure the velocity of a particle in a known centrifugal field. *S* units need to be normalized according to the composition of the medium. Sedimentation coefficients are usually determined with an analytical ultracentrifugation, also referred to as a model E ultracentrifuge. The rotor in the model *E* centrifuge allows the sample to be monitored spectrophotometrically as the sample is centrifuging. As the sample migrates in the centrifugal field the absorbance across the chamber will change. The sedimentation rate can be calculated from this change in absorbance. In addition, physical characteristics (such as size, density and shape) of a particle, or molecule, can be determined from the sedimentation rate of a particle in a medium of known composition.

Relative centrifugal force (RCF) is defined as the ratio of the centrifugal force to the force of gravity:

$$\mathrm{RCF} = F_c/F_g = \omega^2 r/980$$

ω (expressed in radians/sec) is converted to revolutions per minute (rpm) by substituting $\omega = \pi(\mathrm{rpm})/30$, resulting in:

$$\mathrm{RCF} = 1.119 \times 10^{-5}(\mathrm{rpm})^2\, r$$

Where r is expressed in cm. RCF units are expressed as "x g".

Sedimentation of a particular particle is dependent upon the RCF and how long the centrifugal force is applied. Samples must be centrifuged long enough for the particles at the top of the tube to reach the bottom. The size and shape of the centrifuge tube will also affect the centrifugation time. In addition, there is a difference in the RCF between the top and bottom of the tube. These factors need to be taken into consideration when scaling up from small pilot experiments carried out in small centrifuge tubes into large scale experiments. The term g·min refers to the product of the minutes a sample is centrifuged and the g-force. In general, assuming that the sizes and dimensions of the centrifuge tubes are similar, the product of the g-force and the time of centrifugation can be used to determine the conditions needed to completely sediment a particle. For example, if it takes 60 min to completely sediment a particle at 1000 xg, then it should only take 10 min to sediment the same particle at 6000 xg.

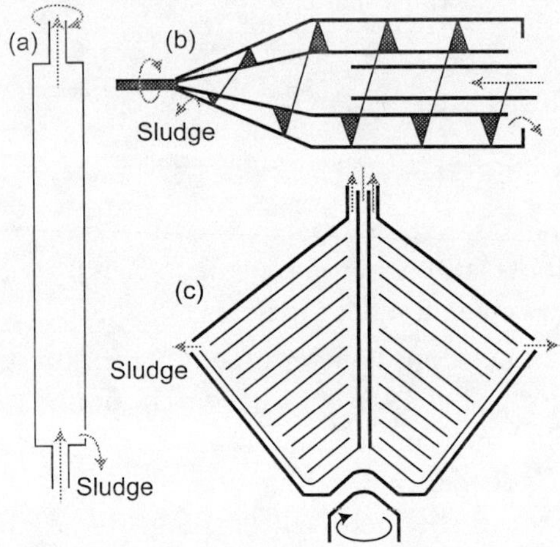

Fig. 4 Basic designs of industrial centrifuges showing the flow of material within the bowls. Motor drives, cooling jackets and sludge collection vessels are not shown. (a) Tubular bowl centrifuge. (b) Continuous scroll centrifuge. (c) Continuous multi chamber disc-stack centrifuge.

4.4 COAGULATION AND FLOCCULATION

All waters, especially surface waters, contain both dissolved and suspended particles. Coagulation and flocculation processes are used to separate the suspended solids portion from the water. The suspended particles vary considerably in source, composition charge, particle size, shape, and density. Correct application of coagulation and flocculation processes and selection of the coagulants depend upon understanding the interaction between these factors. The small particles are stabilized (kept in suspension) by the action of physical forces on the particles themselves. One of the forces playing a dominant role in stabilization results from the surface charge present on the particles. Most solids suspended in water possess a negative charge and, since they have the same type of surface charge, repel each other when they come close together. Therefore, they will remain in suspension rather than clump together and settle out of the water.

The first step destabilizes the particle's charges. Coagulants with charges opposite those of the suspended solids are added to the water to neutralize the negative charges on dispersed non-settable solids such as clay and color-producing organic substances. Once the charge is neutralized, the small suspended particles are capable of sticking together. The slightly larger particles formed through this process and called microflocs, are not visible to the naked eye. The water surrounding the newly formed microflocs should be clear. If it is not, all the particles' charges have not been neutralized, and coagulation has not been carried to completion. More coagulant may need to be added. A high-energy, rapid-mix to properly disperse the coagulant and promote particle collisions is needed to achieve good coagulation. Over-mixing does not affect coagulation, but insufficient mixing will leave this step incomplete. Coagulants should be added where sufficient mixing will occur. Proper contact time in the rapid-mix chamber is typically 1 to 3 minutes.

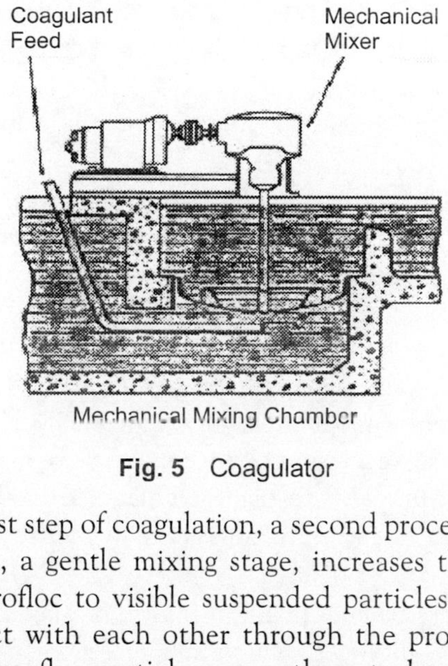

Fig. 5 Coagulator

Following the first step of coagulation, a second process called flocculation occurs. Flocculation, a gentle mixing stage, increases the particle size from submicroscopic microfloc to visible suspended particles. The microflocs are brought into contact with each other through the process of slow mixing. Collisions of the microfloc particles cause them to bond to produce larger, visible flocs called pinflocs. The floc size continues to build through additional collisions and interaction with inorganic polymers formed by the coagulant or with organic polymers added. Macroflocs are formed. High molecular weight polymers, called coagulant aids, may be added during this step to help bridge, bind, and strengthen the floc, add weight, and increase settling rate. Once the floc has reached it optimum size and strength, the water is ready for the sedimentation process.

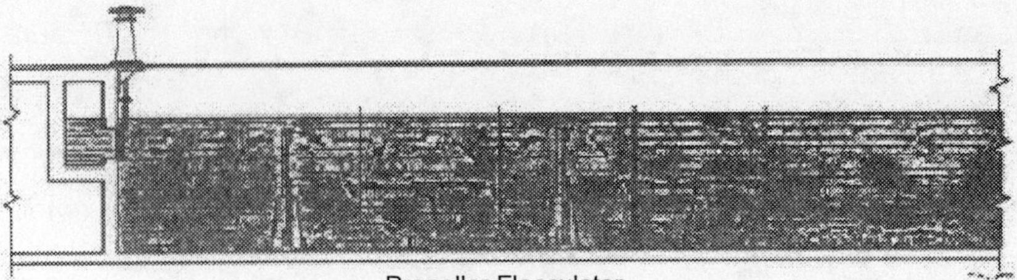

Fig. 6 Flocculator

4.5 CENTRIFUGE EQUIPMENT

A **centrifuge** is a piece of equipment, generally driven by a motor that puts an object in rotation around a fixed axis, applying force perpendicular to the axis. The acceleration achieved by centrifugation is expressed as a multiple of the earth's gravitational force ($g = 9.81\ ms^{-2}$). Bench-top centrifuges can reach acceleration values of up to 15000 *g*, while high-speed refrigerated centrifuges can reach 50000 *g* and ultracentrifuges, which operate with refrigeration and in a vacuum, can reach 500000 *g*. Thus we have abundant range of different classes and kinds of centrifuges, including those for very specialized purposes.

Simple centrifuges are used in biology and biochemistry for isolating and separating biomolecules, cell organelles, or whole cells. They vary widely in speed and capacity. They usually comprise a rotor containing two, four, six, or many more numbered wells within which centrifuge tips may be placed. The rotor is covered by a plastic cover. The cover is usually interlocked to prevent the motor from turning the rotor when it is open, and from allowing the cover to be opened before the rotor stops for several minutes. The cover protects the user from being injured by touching a rapidly spinning rotor. It also protects the user from fragments in case the rotor fails catastrophically. To ensure that the rotor does not turn unbalanced, it must be balanced by placing samples or blanks of equal mass opposite each other. Since most of the mass is derived from the solvent, it is usually sufficient to place blanks or other samples of equal volume. As a safety feature, some centrifuges may stop turning when wobbling is detected.

The centrifuge tips are usually made of plastic or glass; they may vary in capacity from tens of milliliters, to much smaller capacities used in micro centrifuges used extensively in molecular biology laboratories. These micro centrifuges typically accommodate microcentrifuge tubes with capacities from 250 μl to 2.0 ml.

Further, A Centrifuge equipment (or process) needs can be enumerated as following:

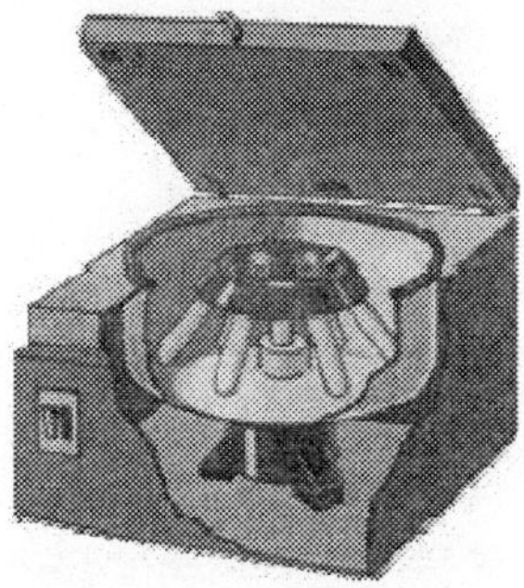

Fig. 7 Centrifuge

1. Cleaning in place (CIP).
2. Steam in place (SIP).
3. Automation for start-up and operation.
4. Short process times.
5. Flexibility in process steam properties.
6. Temperature control.
7. Containment to prevent entry of contaminants and release of solvents or toxic materials.
8. Corrosion resistance.
9. Fail safe system in case of off-balance.
10. Continuous operation for long times.
11. Interface with upstream and downstream operations.

Table 2 illustrates common centrifuge classes and their applications.

Table. 2 Centrifuge classes and their application

	Centrifuge Classes		
	Lowspeed	**High-speed**	**Ultra/micro-ultra**
Maximum Speed (rpm × 10^3)	10	28	100/150
Maximum RCF (× 10^3)	7	100	800/900
Pelleting applications			
Bacteria	Yes	Yes	(Yes)
Animal and plant cells	Yes	Yes	(Yes)
Nuclei	Yes	Yes	(Yes)
Precipitates	Some	Most	(Yes)
Membrane fractions	Some	Some	Yes
Ribosomes/Polysomes	–	–	Yes
Macromolecules	–	–	Yes
Viruses	–	Most	Yes

( ) = can be done but not usually used for this purpose.

4.6 TYPES OF ROTORS

Rotors can be broadly classified into three common categories namely swinging-bucket rotors, fixed-angle rotors, and vertical rotors (Figure 8, Table 3). Note that each type of rotor has strengths and limitations depending on the type of separation. Each category is designed to address three key factors: (i) type of centrifugation (differential, rate-zonal, or isopycnic), (ii) speed, and (iii) volume range.

Of these categories, fixed-angle and swinging-bucket rotors are the most common styles for bench top, low speed, and high-speed floor-model centrifuge applications. Vertical rotors are used primarily in ultracentrifugation. In (1) swinging bucket rotors, the sample tubes are loaded into individual buckets that hang vertically while the rotor is at rest. When the rotor begins to rotate the buckets swing out to a horizontal position (Figure 8). This rotor is particularly useful when samples are to be resolved in density gradients. The longer pathlength permits better separation of individual particle types from a mixture. However, this rotor is relatively inefficient for pelleting. Also, care must be taken to avoid "point loads" caused by spinning CsCl or other dense gradient materials that can precipitate.

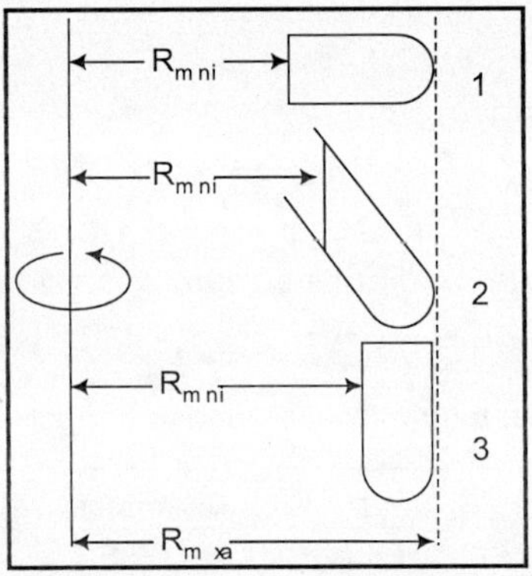

Fig. 8 Rotor types: 1. Swinging bucket rotor, 2. Fixed angle rotor, 3. Vertical rotor.

Table. 3 Types of rotors and their application.

Type of rotor	Pelleting	Rate-zonal Sedimentation	Isopycnic
Fixed-angle	Excellent	Limited	Variable*
Swinging-Bucket	Inefficient	Good	Good**
Vertical	NS	Good	Excellent
Zonal	NS	Excellent	Good

NS = Not suitable.

* Good for macromolecules, poor for cells, and organelles.

** Good for cells and organelles, caution needed if used with CsCl.

In (2) fixed-angle rotors, the sample tubes are held fixed at the angle of the rotor cavity. When the rotor begins to rotate, the solution in the tubes reorients (Figure 8). This rotor type is most commonly used for pelleting applications. Examples include pelleting bacteria, yeast, and other mammalian cells. It is also useful for isopycnic separations of macromolecules such as nucleic acids. In (3) vertical rotors, sample tubes are held in vertical position during rotation. This type of rotor is not suitable for pelleting applications but is most efficient for isopycnic (density) separations due to the short pathlength. Applications include plasmid DNA, RNA, and lipoprotein isolations.

4.6.1 Swinging-bucket Rotors

Swinging-bucket rotors are ideal for separating large-volume samples (up to 12 L) at low speeds. A swinging-bucket rotor system consists of three parts: 1) the rotor body attaches to the centrifuge drive and has four or six arms to support the buckets, 2) the buckets are placed onto the arms of the rotor body, and 3) trunnion pins are used to hold the buckets in place. The above components form the basic swinging-bucket rotor framework; additional accessories can be added as needed to tailor the rotor for a specific application or sample format. For example, large-volume rotors frequently offer a wide variety of adapters (plastic inserts) that can be placed into the buckets to hold the desired tube size. Certain buckets offer sealing lids, which provide bio-containment for potentially hazardous samples.

The most common benchtop and superspeed centrifuge applications requiring a swinging-bucket rotor include: (1) high-throughput protocols such as batch harvesting of whole cells from growth media, (2) high-capacity processing of blood collection tubes, and (3) large-volume tissue culture processing. For these applications, the user can start with volumes as large as 1000 ml and scale down to smaller volumes of 1.5 ml conical tubes. For ultracentrifuge applications, a swinging-bucket rotor typically supports sample volumes ranging from 36 ml to 2.2 ml (adapters are not normally used). An ultracentrifuge equipped with a swinging-bucket rotor can support two types of separations: rate-zonal (i.e., based on mass or size) or isopycnic (i.e., based on density). For rate-zonal separations, a swinging-bucket rotor is advantageous because the path length of the gradient (i.e., distance from R_{min} [inside top of meniscus] to R_{max} [outside bottom of tube]) is long enough for separation to occur. Also, since the buckets are positioned at 90° during the run and returned to a vertical position at the end of the run, the sample retains its orientation, which minimizes any pellet or band disturbance.

4.6.2 Fixed-angle Rotors

Fixed-angle rotors are the most ubiquitous rotors used in centrifugation. The majority are used for basic pelleting applications (differential separations), either to pellet particles from a suspension and discard the excess debris, or to collect the pellet. The cavities in these rotors range in volume from 0.2 mL to 1 L, with speeds ranging from single digits to 1,000,000 xg (relative centrifugal force, RCF). Two factors determine the type of fixed-angle rotor required: desired g-force (RCF) and desired volume. Generally speaking, the size of the rotor is inversely proportional to its maximum speed capability (i.e., the larger the rotor, the lower the maximum speed). An important specification when selecting a fixed-angle rotor is the K factor, which indicates the pelleting efficiency of the rotor at top speed, taking into account the maximum and minimum radius (pathlength) of the rotor cavity. A low K factor indicates a higher pelleting efficiency; therefore, the K factor can be a useful metric for comparing the speed at which particles will pellet across a range of rotors.

4.6.3 Vertical Rotors

Vertical rotors are fairly specialized—their most common use is during ultracentrifugation for isopycnic separations, specifically for the banding of DNA in cesium chloride. In this type of separation, the density range of the solution contains the same density as the particle of interest; thus the particles will orient within this portion of the gradient. Isopycnic separations are not dependent on the path length of the gradient but rather on run time, which must be sufficient for the particles to orient at the proper position within the gradient. Vertical rotors have very low K factors (typically in the range of 5–25), indicating that the particle must only travel a short distance to pellet (or in this case form a band); therefore run time is minimized. Once it is determined that a vertical rotor is appropriate for the end-user application, volume and speed become the deciding factors for which rotor to use.

4.7 TYPES OF CENTRIFUGAL SEPARATIONS

4.7.1 Differential Centrifugation

Separation is achieved primarily based on the size of the particles in differential centrifugation. This type of separation is commonly used in simple pelleting and in obtaining partially-pure preparation of subcellular organelles and macromolecules. For the study of subcellular organelles, tissue or cells are first disrupted to release their internal contents. This crude disrupted cell mixture is referred to as a homogenate. During centrifugation of a cell homogenate, larger particles sediment faster than smaller ones and this provides the basis for obtaining crude organelle fractions by differential centrifugation. A cell homogenate can be centrifuged at a series of progressively higher g-forces and times to generate pellets of partially-purified organelles.

When a cell homogenate is centrifuged at 1000 xg for 10 minutes, unbroken cells and heavy nuclei pellet to the bottom of the tube. The supernatant can be further centrifuged at 10,000 xg for 20 minutes to pellet subcellular organelles of intermediate velocities such as mitochondria, lysosomes, and microbodies. Some of these sedimenting organelles can be obtained in partial purity and are typically contaminated with other particles. Repeated washing of the pellets by resuspending in isotonic solvents and re-pelleting may result in removal of contaminants that are smaller in size (Figure 9). Obtaining partially-purified organelles by differential centrifugation serves as the preliminary step for further purification using other types of centrifugal separation (density gradient separation).

Usually use a fixed angle rotor.

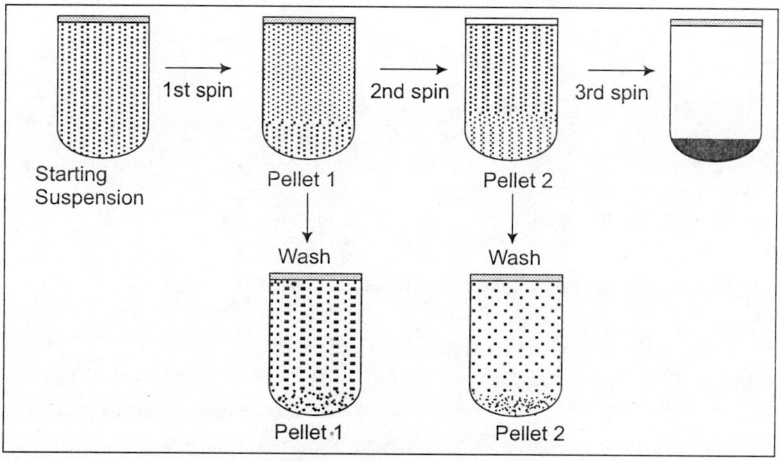

Fig. 9 Differential centrifugation

4.7.2 Density Gradient Centrifugation

Density gradient centrifugation is the preferred method to purify subcellular organelles and macromolecules. Density gradients can be generated by placing layer after layer of gradient media such as sucrose in a tube with the heaviest layer at the bottom and the lightest at the top in either a discontinuous or continuous mode. The cell fraction to be separated is placed on top of the layer and centrifuged. Density gradient separation can be classified into two categories. 2a). Rate-Zonal (size) separation. 2b). Isopycnic (density) separation.

4.7.2.1 Rate zonal (size) separation

Rate-Zonal separation takes advantage of particle size and mass instead of particle density for sedimentation. Fig. 10 illustrates a rate-zonal separation process and the criteria for successful rate-zonal separation. Examples of common applications include separation of cellular organelles such as

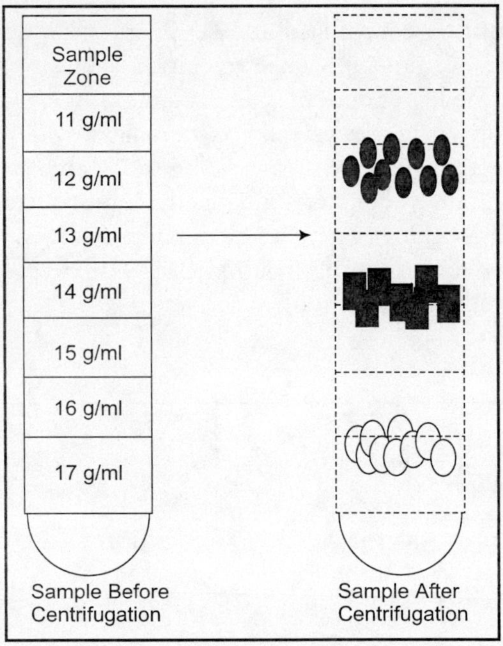

Fig. 10 Rate zonal (size) separation.

endosomes or separation of proteins, such as antibodies. For instance, Antibody classes all have very similar densities, but different masses. Thus, separation based on mass will separate the different classes, whereas separation based on density will not be able to resolve these antibody classes. Swinging bucket rotor always used.

Criteria for successful rate-zonal centrifugation:

- Density of the sample solution must be less than that of the lowest density portion of the gradient.
- Density of the sample particle must be greater than that of the highest density portion of the gradient.
- The pathlength of the gradient must be sufficient for the separation to occur.
- Time is important. If you perform too long runs, particles may all pellet at the bottom of the tube.

4.7.2.2 Isopycnic separation

In this type of separation, a particle of a particular density will sink during centrifugation until a position is reached where the density of the surrounding solution is exactly the same as the density of the particle. Once this quasi-equilibrium is reached, the length of centrifugation does not have any influence

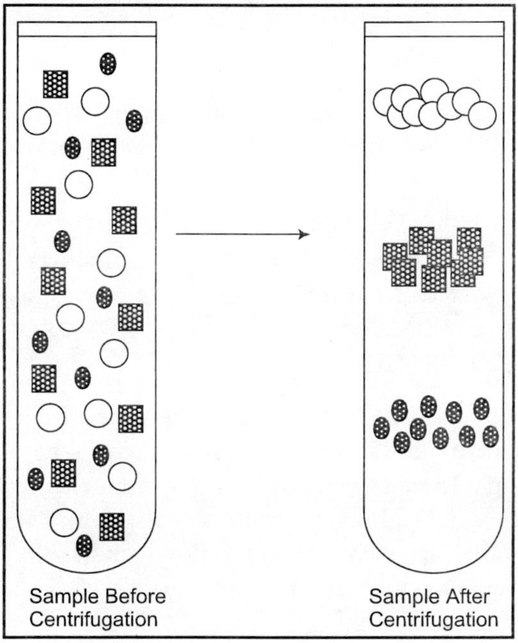

Fig. 11 Isopycnic separation.

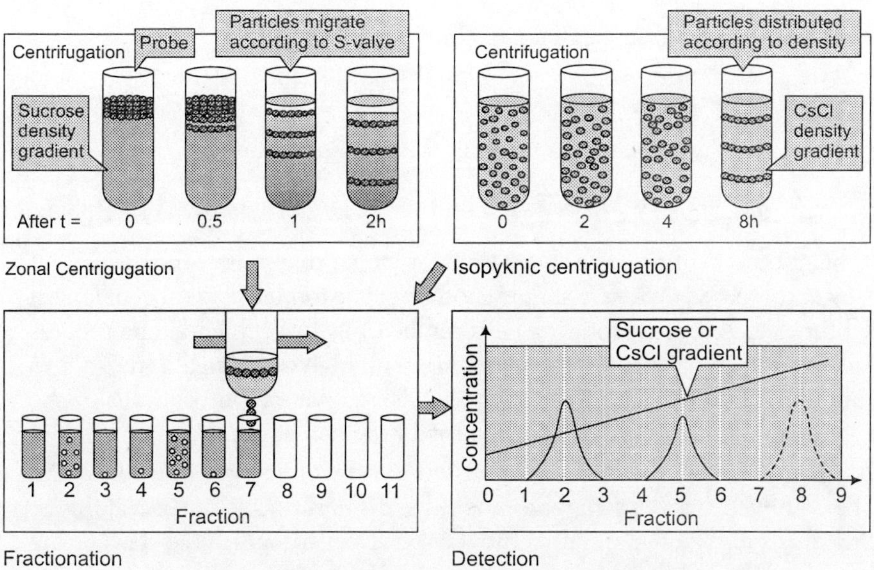

Fig. 12 Overview of density gradient centrifugation

on the migration of the particle. A common example for this method is separation of nucleic acids in a CsCl gradient. Can use swinging bucket or fixed angle rotor.

Criteria for successful isopycnic separation:

- Density of the sample particle must fall within the limits of the gradient densities.
- Any gradient length is acceptable.
- The run time must be sufficient for the particles to band at their isopycnic point. Excessive run times have no adverse effect.

4.8 TYPES OF CENTRIFUGES

Depending on a particular application centrifuges differ in their sizes and design. A common part in all the centrifuges is the central motor that spins a rotor having the sample to be separated. Particles of biochemical interest are usually suspended in a liquid buffer system contained in a specific tube or separation chambers that are located in specialized rotors. The optimum pH value, stability factor, salt concentrations have to be carefully evaluated in order to preserve biological function.

The major differences in centrifuges are:

- The maximum speed at which the biological samples are subjected to increased sedimentation.
- Te presence or absence of a vacuum.
- The maximum volume of samples and capacity for individual centrifugation tubes.

Centrifuges may be classified into following major groups:

4.8.1 Tubular Bowl Centrifuge

The tubular bowl centrifuge has been used for longer than most other designs of centrifuge. It is based on a very simple geometry: it is formed by a tube, of length several times its diameter, rotating between bearings at each end. The process stream enters at the bottom of the centrifuge and high centrifugal forces act to separate out the solids. The bulk of the solids will adhere on the walls of the bowl, while the liquid phase exits at the top of the centrifuge.

As this type of system lacks a provision of solids rejection, the solids can only be removed by stopping the machine, dismantling it and scraping or flushing the solids out manually. Tubular bowl centrifuges have dewatering capacity, but limited solids capacity. Foaming can be a problem unless the system includes special skimming or centripetal pumps.

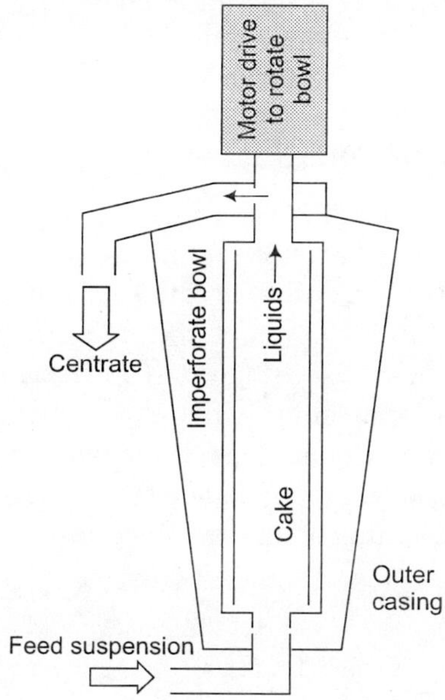

Fig. 13 Tubular bowl centrifuge

4.8.2 Chamber Bowl Centrifuge

The chamber bowl centrifuge is a number of tubular bowls arranged co-axially. It has a main bowl containing cylindrical inserts that divide the volume of the bowl into a series of annular chambers, which operate in series. Feed enters the center of the bowl and the suspension passes through each chamber in turn, at increasing distances from the axis. The solids settle onto the outer wall of each chamber and the clarified liquid emerges as an overflow from the largest diameter chamber. This device provides also a classification of the suspended

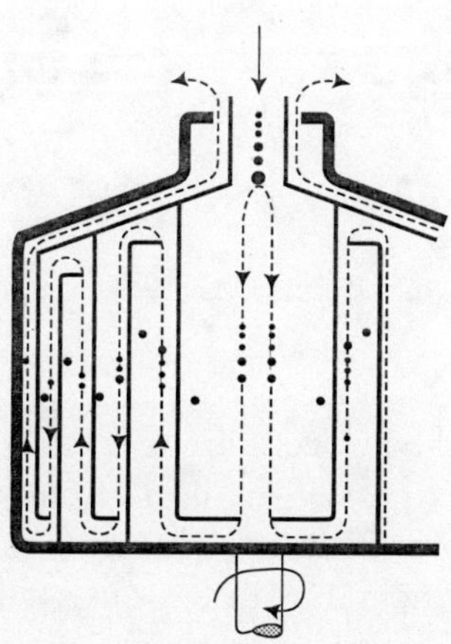

Fig. 14 Chamber bowl centrifuge

solids: the coarse particles deposit in the inner chamber and the increasingly fine particle deposit on the subsequent chambers. The removal of sedimented solids requires the stopping of rotation for manual cleaning.

4.8.3 Imperforate Basket Centrifuge

The imperforate basket centrifuge is used when if the solid content of the suspension is higher. It consists of a simple drum-shaped basket or bowl, usually rotating around a vertical axis. The solids accumulate and compress as effect of the centrifugal force, but they are not dewatered. The residual liquid is drained out when the rotation of the bowl is stopped. The layer of solids is removed manually by scraping or shoveling. Unloading can be achieved semi automatically first by use of a skimmer pipe to remove the residual liquid and then by lowering a knife blade into the solid and so cutting it out from the bowl. This allows avoiding the switching off of the machine.

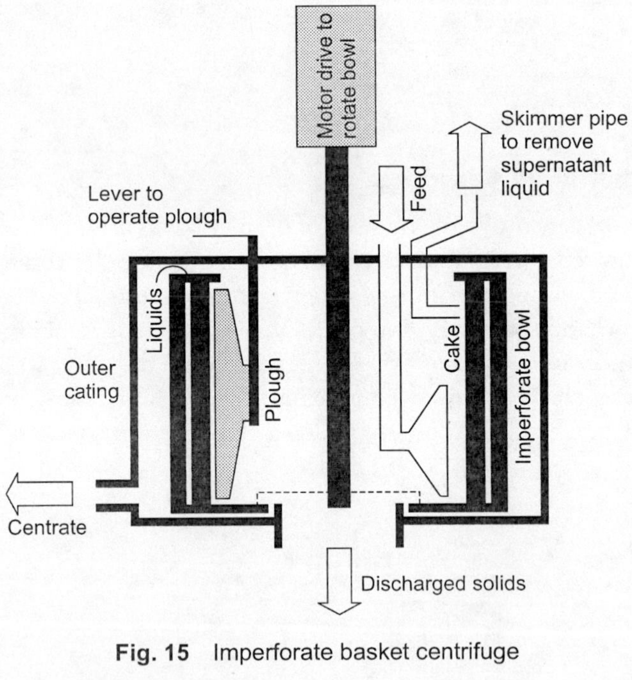

Fig. 15 Imperforate basket centrifuge

4.8.4 Disk Stack Separator

The simplest design is a closed bowl, containing the disk stack, with any solids present collecting at the outer part of the bowl, from which they have to be removed manually after stopping rotation. The solids are discharged from the bowl by a number of methods, including the basic use of nozzles, which are open continuously, allowing thick slurry to discharge. In the more complicated

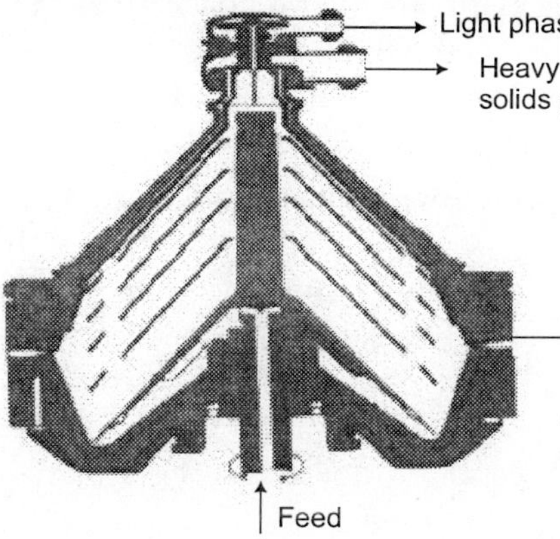

Fig. 16 Disk stack separator

design valved nozzles open automatically when the solid depth in the bowl reaches a certain value, and then close again when most of the solids have been discharged. In the most complicated design the bowl is opened: its shell splits circumferentially for a short period, with the opening also controlled by solids depth in the bowl.

4.8.5 Decanter

The decanter centrifuge is the only sedimentation centrifuge designed from the start to handle significant solid concentration in the feed suspension. At the same time it can achieve quite good degrees of clarification of the liquid concentrate. Although a complicated piece of machinery it embodies a simple principle. They consist basically of a horizontal cylindrical bowl rotating at a high speed, with a helical extraction screw placed coaxially. The screw perfectly fits the internal contour of the bowl, only allowing clearance between the bowl and the scroll. The differential speed between screw and scroll provides the conveying motion to collect and remove the solids, which accumulate at the bowl wall.

The product to be treated is introduced axially into the unit by appropriate distributor. It is propelled into the ring space formed by the internal surface of the bowl and the body of the scroll. The separation process basically takes place inside the cylindrical section of the bowl. The relative velocity of the scroll pushes the settled product along into the bowl. The conveyance of the solids into the length of the cone enables the sediment to pass out of the clarified liquid phase. As the feed is continuous a liquid level is established in the unit following a cylindrical surface that constitutes the internal surface of the liquid ring. Once the solids have passed out of the liquid ring the remaining section of the cone all the way up to the ejector provide the final draining: this section is known as the drying zone. The clarified liquid is collected at the other end of the bowl by flowing through the adjustable threshold, which restrict the liquid ring of the unit. A cover that enables the clarified liquid as well as the sediments to be collected protects the rotor.

The decanter operates mainly by sedimentation a process causing the separation of suspended solids by virtue of their higher density than the liquid in which they are suspended. If the density difference is high than gravity may provide sufficient driving force for the separation to occur in a reasonable time. If the density difference is small, or the particle size is very small, than gravity separation would take too long and the separation force must be increased by the imposition of centrifugal forces many times that of gravity alone.

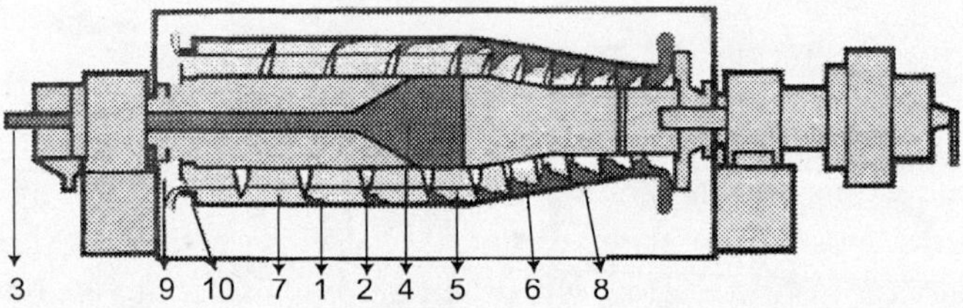

Fig. 17 Decanter and its parts. 1. Cylindroconical bowl. 2. Helical extraction screw (scroll). 3. Feed. 4. Distributor. 5. Ring space. 6. Settled product. 7. Liquid level. 8. Drying zone. 9. Clarified liquid. 10. Adjustable thresholds.

Density centrifugation uses a dense solution or density gradient to separate particles based on their individual densities or mass/size ratio. A solution is prepared such that a gradient of densities is available for particles to pass through or float upon. This density gradient may be continuous or prepared in a stepped manner such as the layers in a parfait. For instance, sucrose is often used to prepare density gradients and one can carefully float a solution of 40% sucrose onto a layer of 45% sucrose and add further less dense layers above. The homogenate, prepared in a dilute buffer and centrifuged briefly to remove tissue and unbroken cells, is then layered on top. After centrifugation typically for an hour at about 100,000 xg, one can observe disks of cellular components residing at the change in density from one layer to the next. By carefully adjusting the layer densities to match the cell type, one can enrich nuclear, mitochondrial, microsomal, and plasma membrane vesicle fractions. In another common practice, cesium chloride is mixed with a buffer and continuous density gradient forms during the centrifugation. Nucleic acids such as plasmid and genomic DNA as well as RNA appear as floating clouds in this gradient after centrifugation and can be extracted for further purposes. This procedure is also known as "equilibrium density centrifugation".

4.9 SELECTION OF CENTRIFUGE TUBES

Table 4 and Table 5 illustrate properties of centrifuge tubes and the proper rotors in which they should be used.

Table 4 Chemical compatibility of Popular Tube Materials

Tube Plastic type Clarity	Chemical	Chemical Resistance
Polypropylene (PP)	Opaque	Good
Polyallomer (PA)	Opaque	Good
Polycarbonate (PC)	Clear	Poor
Polyethylene terephatalte (PET)	Clear	Poor

Table 5 Tube and Rotor Compatibility

	Rotor Type		
Tube type	Fixed-angle	Swinging-bucket	Vertical
Thin wall open top	No	Yes	No
Thick wall open top	Yes	Yes	No
Thin wall sealed	Yes	Some tube types	Yes
Oak ridge	Yes	No	No

Selection of the appropriate centrifuge tube

- Prevents sample leakage or loss
- Ensures chemical compatibility
- Allows easy sample recovery

Major factor in selection of a tube (plastic) material

- Clarity
- Chemical resistance
- Sealing mechanism (if needed)
- Check product guide pages or tube packaging for notes on recommended sample volume and maximum speed.
- Always run thin-walled, sealed tubes full in a fixed angle or vertical rotor.

Examples:

- open top tube with multiple sealing assembly
- Re-seal tubes
- Ultracrimp® and Clearcrimp® tubes
- Autoclave tubes only if absolutely necessary and only at 121°C for 20 min.
- Avoid cleaning plastic tubes in automated dishwashers or glassware washers, which may produce excessively hot temperatures.

- Clean tubes with a mild laboratory detergent in warm water, rinse, and thin air dry are recommended.

Tube must be carefully matched with rotor type to prevent sample loss and/ or failure as illustrated in Table 5 below, to prolong tube life and avoid breakage or collapse:

4.10 COMMON CENTRIFUGATION VOCABULARY

- Pellet: hard-packed concentration of particles in a tube or rotor after centrifugation.
- Supernatant: The clarified liquid above the pellet.
- Adapter: A device used to fit smaller tubes or centrifugal devices in the rotor cavities.
- RPM: Revolutions Per Minute (Speed).
- Rmax: Maximum radius from the axis of rotation in centimeters.
- Rmin: Minimum radius from the axis of rotation in centimeters.
- RCF: Relative centrifugal Force. RCF = $11.17 \times R_{max}\,(RPM/1000)^2$
- K-factor: Pelleting efficiency of a rotor. Smaller the K-factor, better the pelleting efficiency.

$$K = \frac{2.53 \times 10^{11}\,Ln(R_{max}/R_{min})}{(RPM)^2}$$

- S-value: the sedimentation coefficient is a number that gives information about the molecular weight and shape of the particle. S-value is expressed in Svedberg units. The larger the S-value, the faster the particle separates.
  - *Pelleting time:* time taken to pellet a given particle. T = K/S where T= pellet time in hours. K = K-factor of the rotor, and S = sedimentation coefficient.
  - *Rotor conversion formula:* If the time to pellet a sample in your "old" rotor is known, one could determine the time it would take for the same sample to pellet in a "new" rotor. The formula for this determination is as follows:

$$\frac{T_1}{K_1} = \frac{T_2}{K_2} \rightarrow T_1 = T_2\left(\frac{K_1}{K_2}\right)$$

Where:

T1 = Time to pellet in the "new" rotor

T2 = Time to pellet in the "old" rotor

K1 = K-factor of the "new" rotor

K2 = K-factor of the "old" rotor

5

CHAPTER

PRODUCT ISOLATION

5.1 INTRODUCTION

Separation and isolation of desired product, which can be one or more components, from a complex mixture is a requirement for many operations in the biotechnology industries. The components in question range from particulate materials down to small molecules. The separations usually aim to achieve removal of specific components, in order to increase the added value of the products, which may be the residue, the extracted components or both. All separations rely on exploiting differences in physical or chemical properties of the mixture of components. Some of the more common properties involved in separation processes are particle or molecular size and shape, density, solubility and electrostatic charge. In some operations, more than one of these properties are involved. However, most of the processes involved are of a physical nature.

Separation from solids or liquids involves the transfer of selected components across the boundary of the target product. In many processes another stream or phase is involved, for example in extraction processes. However, this is not always so, for example expression, centrifugation or filtration. In centrifugation, fat can be separated from water due to their density differences, by the application of a centrifugal force. In filtration there is a physical barrier to the transfer of certain components and the liquid is forced through the barrier by pressure, whilst the solids are retained. The resistance to flow will change throughout the filtration process, due to solids build-up. It can be seen that main driving forces in these applications are pressure and density differences. As for all processes, separation rates are very important and these are affected by the size of the driving forces involved.

In situations where a second phase or stream is involved, mass-transfer considerations become important; these involve the transfer of components within the food to the boundary, the transfer across the boundary and into the bulk of the extraction solvent. It is also important to increase the interfacial area exposed to the solvent. Therefore, size reduction, interfacial phenomena, turbulence and diffusivities all play a role in these processes. In many applications this additional stream is a liquid, either water or an organic solvent; more recently supercritical fluids, such as carbon dioxide, have been investigated. However, in hot-air drying the other phase is hot air, which supplies the energy and removes the water. Mass-transfer considerations are important also in some membrane applications and adsorption processes, where the additional stream is a solid. Other examples of driving force are concentration differences and chemical potential, which are involved in these operations.

Two-phase extraction is a useful unit operation both for biologics and small molecules. If the product is intracellular (inside the cell), harvested cells are sometimes pumped directly into an extraction liquid, where they undergo lysis caused by the organic solvents. This method is most suitable for small molecule products. Two-phase extraction, when applied correctly, also works as a concentration method. Very few things are measured in an extraction process (temperature and pH) and the set-points are usually decided before the process.

In some processes, both heat and mass transfer processes are involved. This is especially so for separation reactions involving a change of phase, such as evaporation or sublimation. Heat is required to cause vaporization for evaporation, dehydration and distillation processes. Water has a much higher latent heat of vaporization (2257 kJ/kg) than most other organic solvents. With solid foods the rate of heat transfer through the food may limit the overall process; for example in freeze-drying the process is usually limited by rate of heat transfer through the dry layer.

5.2 THEORY AND BASIC PRINCIPLE

Separation processes may be batch or continuous. A single separation process, for example a batch extraction, involves the contact of the solvent with the food. Initially concentration gradients are high and the rate of extraction is also high. The extraction rate falls exponentially and eventually an equilibrium state is achieved when the rate becomes zero. The extraction process may be accelerated by size reduction, inducing turbulence and increasing the extraction temperature. Equilibrium is achieved either when all the material has been extracted, in situations where the volume of solvent is well in excess of the solute or when the solvent becomes saturated with the solute, i.e. when the solubility limit has been achieved, when there is an excess of solute over the solvent. However, the attainment of equilibrium may take some considerable time. Batch reactions may operate far away from equilibrium or close to it.

Equilibrium data is important in that it provides information on the best conditions that can be achieved at the prevailing conditions. Equilibrium data is usually determined at fixed conditions of temperature and pressure. Some important types of equilibrium data are:

- solubility data for extraction processes;
- vapour/liquid equilibrium data for fractional distillation;
- partition data for selective extraction processes;
- water sorption data for drying.

Continuous processes may be single-stage or multiple-stage processes. The stages themselves may be discrete entities, for example a series of stirred tank reactors, or there may be many stages built into one unit of equipment, for example a distillation column or a screw extractor. The flow of the two streams can either be co-current or counter-current; although counter-current is normally favored as it tends to give a more uniform driving force over the length of the reactor as well as a higher average driving force over the reactor. In some instances a combination of co-current and counter-current may be used; for example in hot air drying, the initial process is co-current to take advantage of the high initial driving rates, whereas the final drying is counter-current to permit drying to a lower moisture content.

Continuous equipment usually operates under steady state conditions, i.e. the driving force changes over the length of the equipment, but at any particular location it remains constant with time. However, when the equipment is first started, it may take some time to achieve steady-state. During this transition period it is said to be operating under unsteady state conditions. In continuous processes the flow may be either streamline or turbulent. Consideration should be taken of residence times and distribution of residence times within the separation process; the two extremes of behavior are plug flow, with no distribution of residence times, through to a well-mixed situation, with an infinite distribution of residence times.

How close the process operates to equilibrium depends upon the operating conditions, flow rates of the two phases, time and temperature. These conditions affect the efficiency of the process, such as the recovery and the size of equipment required.

Finally, all rates of reaction are temperature dependent. Physical processes are no exception, although activation energies are usually much lower than for chemical reaction rates. Using higher temperatures normally increases separation rates. However, there are limitations with biological materials: higher temperatures increase degradation reactions, causing color and flavor changes, enzyme inactivation, protein denaturation, loss of, functionality and a reduction in nutritional value. Safety issues with respect to microbial growth may also need to be considered.

Extraction of components from liquids can be achieved by contacting the liquid with a solvent which will preferentially absorb the components of interest and can then be separated from the liquid, for example by centrifugation. Such solvent extractions could be used for recovering oils and oil-soluble components of flavor components from liquids.

5.3 EXTRACTION IN GENERAL

Extraction is a process that separates components based upon chemical differences rather than differences in physical properties. The basic principle behind extraction involves the contacting of a solution with another solvent that is immiscible with the original. The solvent is also soluble with a specific solute contained in the solution. Two phases are formed after the addition of the solvent, due to the differences in densities. The solvent is chosen so that the solute in the solution has more affinity toward the added solvent. Therefore mass transfer of the solute from the solution to the solvent occurs. Further separation of the extracted solute and the solvent will be necessary. However, these separation costs may be desirable in contrast to distillation and other separation processes for situations where extraction is applicable.

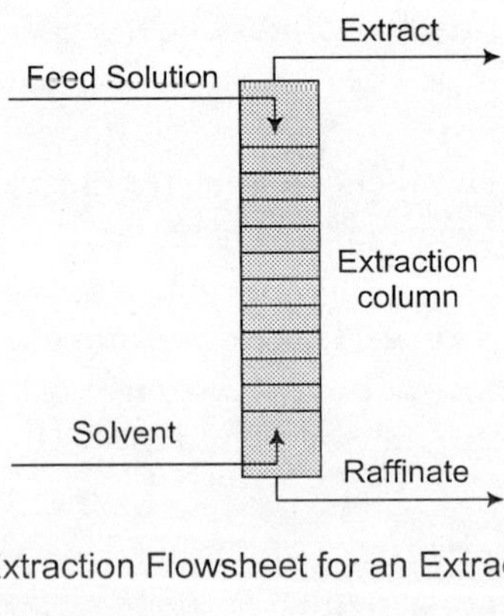

Fig. 5.1 Extraction Flowsheet for an Extractor Column

A general extraction column has two input stream and two output streams. The input streams consist of a solution feed at the top containing the solute to be extracted and a solvent feed at the bottom which extracts the solute from the solution. The solvent containing the extracted solute leaves the top of the column and is referred to as the extract stream. The solution exits the bottom of the column containing only small amounts of solute and is known as the raffinate. Further separation of the output streams may be required through other separation processes.

5.3.1 Process Operation

There are certain design variables that must be assigned in an extraction process.

- Operating Temperature
- Operating Pressure
- Feed Flow Rate
    - o Composition
    - o Temperature of entering stream
    - o Pressure of entering stream

As in many separation processes, the pressure and temperature conditions play a large role in the effectiveness of the separation. In order for a good split of the feed the pressure and temperature must be such so as to ensure that all components remain in the liquid phase. The process will be adversely affected if one or more of the components are allowed to become a vapor, or the extraction may not occur at all if a large enough portion of a component is allowed to vaporize. In addition, the temperature should be high enough that the components are all soluble with one another. If extremes in temperature are present, finding a suitable solvent for extraction can be problematic. This is however generally not the case since one of the biggest benefits in the extraction process is that it can be done at ambient pressures and temperatures.

In many applications, a separation process is desired where an extreme temperature will destroy the desired product such as the pharmaceutical industry. For these applications, extraction is ideally suited, since the only temperature requirement is that dictated by the solubility. At this point the biggest challenge would be finding a suitable solvent for the extraction.

We can also use the pharmaceutical industry in another example for the benefits of extraction and this has to do with the volumes involved for effective extraction. The extraction process can become very expensive if the solvent needed to be used is costly these expenses can be contained if a batch process is being used and this is often the case in medicines. In a non-batch process the solvent would need to be constantly supplied and this would involve either a huge amount of solvent or another separation process in order to recycle the solvent.

5.3.2 Limitations

We must consider the under what extremes extraction can be used as separation process.

1. Suitable Solvent
    - Solvent partially soluble with the carrier.
    - Feed components immiscible with the solvent.

- Solute is soluble in the carrier and at the same time completely or partially soluble in the solvent.
- Different densities than the feed components for a phase separation to facilitate and maintain the capacity of the extractor high.
- Extremely high selectivity for the solute for the solvent to dissolve the maximum amount of solute and the minimum amount of the carrier.
- Large distribution coefficient to reduce the theoretical number of stages contributing to a greater efficiency.
- Low viscosity increases the capacity of the extraction column and does not allow for the settling rate of dispersion to be slow.
- Chemically stable and inert toward other components of the system.
- Low cost, nontoxic, and nonflammable

2. Equipment
   - Interfacial tension and Viscosity.
   - High interfacial tension and viscosity leads to more power being supplied to maintain rapid mass transfer throughout the extraction process.
   - Low interfacial tension and viscosity leads to the formation of an emulsion.
3. Temperature preferred to be higher since solubility increases, but temperature not higher than the critical solution temperature.
4. Pressure for condensed system must be maintained below the vapor pressure of the solutions such that a vapor phase will not appear and interrupt liquid equilibrium.
5. Separation may only occur for compositions in the region between the feed composition and that apex of the carrier.

5.3.3 Applicability

With all the key components in the design of an extractor system to be discussed, the equipment selection can be evaluated. We must determine which extractor would apply for the situation at hand. The specifications for each of these different systems are relatively the same. The following design constraints should be placed on each system in order to optimize the individual process: (i) maximize surface area of mass transfer, and (ii) adjust flow feeds for maximum solute recovery.

In general, there are three main types of extractors to focus on:

(1) **Mixer-settlers:** Mixer-settlers are used when there will only be one equilibrium stage in the process. For such a system, the two liquid phases

are added and mixed. Due to their density differences, one phase will settle out and the mixture will be separated. The downfall to this type of extractor is that it requires a large-volume vessel and a high liquid demand.

(2) **Contacting columns:** Contacting columns are practical for most liquid-liquid extraction systems. The packing, trays, or sprays increase the surface area in which the two liquid phases can intermingle. This also allows for a longer flow path that the solution can travel through. In the selection of a packing, it is necessary to select a material that is wetted by the continuous phase. Lastly, the flow in a column should always be counter-current.

(3) **Centrifugal contractors:** Centrifugal contractors are ideal for systems in which the density difference is less than 4%. In addition, this type of system should be utilized if process requires many equilibrium stages. In these systems, mechanical devices are used to agitate the mixture to increase the interfacial area and decrease mass transfer resistance.

Many different types of centrifugal contractors exist, and each has its own guidelines for operation and selection. A more detailed view can be found in Section 5.5.

5.3.4 Mode of Operation

Extractors can be operated in cross-current or counter-current mode. The following section compares these configurations.

(i) Cross-current Operation

Cross-current mode is mostly used in batch operation. Batch extractors have traditionally been used in low capacity multi-product plants such as are typical in the pharmaceutical and agrochemical industries. For washing and neutralization operations that require very few stages, cross-current operation is particularly practical and economical and offers a great deal of flexibility. The extraction equipment is usually an agitated tank that may also be used for the reaction steps. In these tanks, solvent is first added to the feed; the contents are mixed, settled and then separated. Single stage extraction is used when the extraction is fairly simple and can be achieved without a high amount of solvent. If more than one stage is required, multiple solvent-washes are given.

Though operation in cross-current mode offers more flexibility, it is not very desirable due to the high solvent requirements and low extraction yields.

(ii) Counter-current Operation

As described above, the cross-current operation is mostly used in low capacity multi-product batch plants. For larger volume operation and more efficient use of solvent, counter-current mixer-settlers or columns are employed. Counter-current operation conserves the mass transfer driving force and hence gives optimal performance.

The extraction factor (E), is an important parameter in the design of extraction processes. For a given number of stages, the higher the E factor, the higher is the reduction ratio and easier is the extraction. Systems with E of less than 1.3 are not likely to be commercially feasible.

5.3.5 General Flow Sheet of Extraction Process

With the key components of extraction discussed, the following general flowchart can be utilized for almost any process. In this process, depicted in Figure 5.2., we will assume that the feed (F) contains components A and C. A solvent (S) is introduced in such a way that it will extract C from the feed. The raffinate composition (R) is specified with respect to the recovery of C that is needed.

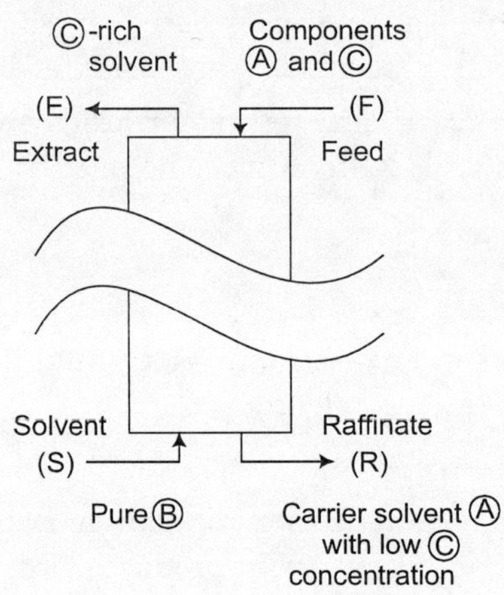

Fig. 5.2 A general extraction process diagram.

5.4 SUPER CRITICAL FLUID EXTRACTION

5.4.1 Introduction

Solvent extraction is one of the oldest methods of separation known and certainly dates back to prehistory. The science of solvent extraction has evolved accordingly over a long period of time and much progress has been made in the understanding of solvation and the properties of liquid mixtures used in extraction processes. The associated literature on phase behavior is certainly extensive and, although representation of highly non-ideal mixtures is still problematic, many theoretical models have been successfully developed. Extensive databanks of pure component properties have grown to support such models in order to predict solvent performance in process applications. Today, even with the introduction of new separation technologies, solvent extraction remains one of the most widespread techniques operating on an industrial scale. Hannay and Hogarth's (1879) early observations of the dissolution of solutes in supercritical fluid (SCF) media introduced the possibility of a new solvent medium. However, it is only in recent years (since 1960) that commercial process applications of supercritical fluid extraction have been extensively examined.

In the last decade many advances have been made in researching SCF extraction both in terms of fundamental aspects and commercial applications. In particular the high degree of selectivity and control over solubilities afforded by pressure (and temperature) variation has led to the introduction of many novel SCF extraction and fractionation processes. Of all possible gases, the benign properties (non-toxic, non-flammable) and accessible critical temperature of CO_2 have ensured its predominance as a safe SCF solvent for the food industry. The essential features of a modern solvent extraction process (using a liquid or SCF solvent medium) are illustrated schematically in Figure 5.3. The material to be extracted is placed in an extraction vessel (extractor) into which solvent is introduced under conditions (temperature, flow rate etc.) which optimize the dissolution of the desired components. The solvent stream is then passed to a separation vessel (separator) where conditions are set to selectively separate the solvent from the extracted components. The solvent is then condensed and recycled through the system.

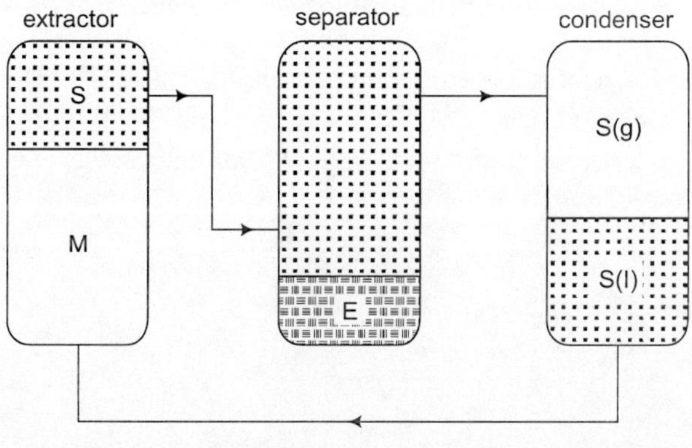

Fig. 5.3 Schematic representation of a solvent extraction process (S = solvent; M = material extracted; E = extract)

In conventional liquid extraction, solvents of low volatility are employed with vapor pressures less than one atmosphere. In the course of the extraction process the solvent exists as a liquid during the extraction stage and a gas when it is removed from the extract by distillation. Variations in pressure are small and do not significantly exceed the vapor pressure of the liquid at the extraction temperature. Although temperature variation gives some control over solubility, selective removal of components from a mixture is largely determined by the chemical nature of the solvent. Progressive fractionation can therefore only be achieved by a fortuitous response to temperature or by systematically changing the solvent or the composition of a mixed-solvent system.

The initial aims of this chapter are to establish the basic principles involved in SCF extraction. Selected applications are later reviewed with reference to the underlying fundamental properties that serve to differentiate the behavior of SCFs from conventional liquid solvents.

5.4.2 The Supercritical Fluid State

The *P-T* phase diagram for CO_2 showing all four physical states (solid, liquid, gas and SCF) is shown in Fig. 5.4. Below the freezing point solid CO_2 ('dry ice') exists, which melts on heating when the thermal energy of the molecules overcomes the lattice energy. The integrity of the liquid state so formed is maintained by relatively weak attractive intermolecular forces (van der Waals). The formation of a supercritical fluid state above the critical temperature (T_c = 31.05 degree C) can be viewed as an analogous process in which the thermal energy of the molecules overcomes all attractive interactions maintaining the liquid state. Like a gas, the SCF state formed then occupies all available volume. Strictly the SCF state exists above both the critical temperature and pressure ($T > T_c$; $P > P_c$) though the latter condition is often relaxed in the technical literature. A substance above its critical temperature therefore behaves like a gas and always occupies all available volume as a single phase. However, unlike a gas, a SCF cannot be condensed to a coexisting liquid-gas state by application of pressure. Similarly when the critical pressure is exceeded it is possible to go from a SCF state to a compressed liquid condition by cooling, but a single-phase filling all available volume is always maintained.

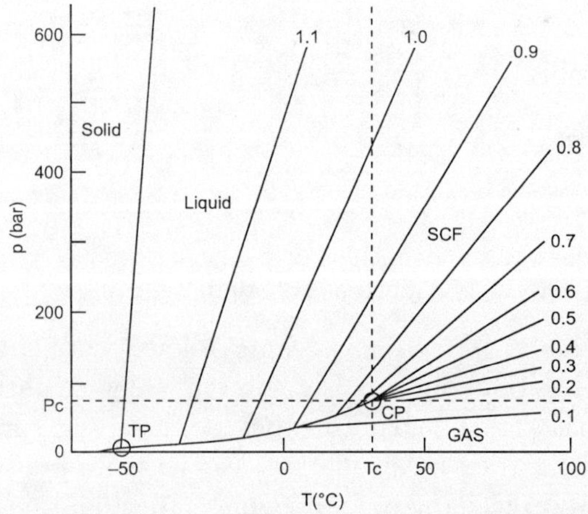

Fig. 5.4 Pressure-temperature phase diagram for CO_2 showing isochores (g cm^{-3}).

It should be appreciated that there are no phase boundaries delineating the SCF state and therefore no sharp changes in physical properties occur on entering this region. Transition to the SCF state from a gas or liquid is thus an 'invisible' process. However, if a coexisting liquid-gas mixture is heated at constant volume along the vapor pressure curve, the density of the liquid phase decreases while that of the gas phase increases, until at the critical point they become equal and the meniscus between them disappears. As this point is approached density fluctuations of microscopic dimensions give rise to a distinctive light-scattering phenomenon known as 'critical opalescence'.

Although the supercritical state offers a greater range of density, which in turn provides greater control over solubilities, the liquid state of compressed gases is often employed in extraction processes, particularly for separation of thermo-labile components at low temperatures. In order to avoid restrictive and confusing nomenclature, it is convenient to use the term 'near-critical liquid' (NCL) to distinguish the state of a compressed gas just below T_c from a 'normal' liquid at NTP, for which $T < T_c$. The term 'near critical fluid' (NCF) will be used in this chapter to represent both SCF and NCL states of compressed-gas solvents.

Many liquids commonly employed as solvents enter an SCF state on heating, but for most purposes the critical temperatures are too high to permit their use as SCF solvents (e.g. T_c for hexane is 234°C). All substances with accessible critical temperatures are gases at NTP and representative examples for use in extraction processes are shown in Table 5.1. Being non-toxic, non-flammable, and chemically inert, CO_2 has obvious practical advantages over other potential gases for use in large-scale extraction processes under pressure.

Table 5.1 Potential gases for near-critical fluid extraction

Name	*Formula*	*T_c (degree C)*	*P_c (bar)*
Carbon dioxide	CO_2	31.1	73.8
Nitrous oxide	N_2O	36.4	71.5
Ammonia	NH_3	132.4	111.3
Ethane	C_2H_6	32.2	48.2
Propane	C_3H_8	96.6	41.9
Ethylene	C_2H_4	9.2	49.7
Freon 13	$CClF_3$	28.9	38.7

5.4.3 Properties of Ncf Solutions

5.4.3.1 Solubilities in Ncfs

There has been much confusion in some of the literature concerning the solvent properties of NCF CO_2. An impression is often given that NCFs are universal solvents which can be 'tuned' to extract virtually any component of a mixture by selecting a suitable set of conditions of temperature and pressure. Statements to the effect that NCFs are 'good' solvents, implying that solute loadings are high, are also prevalent and highly misleading. Before examining the solvent properties of NCFs in detail, it is worth stating a few basic principles:

(1) To be 'supercritical' intermolecular attractive interactions must be relatively weak compared with thermal energy. This necessitates an absence of all polar interactions, such as hydrogen bonding, and defines a medium of low dielectric constant. All NCFs are therefore essentially apolar solvents.

(2) The absence of strong attractive interactions between molecules means that solvation energies are generally low and solubilities in NCFs are thus often much lower than in liquid solvents.

(3) NCFs can be highly discriminating and frequently offer a greater selectivity than liquid solvents. Any attempt at increasing solubilities by changing conditions or injecting entrainers usually serves to reduce selectivity. Although some selectivity is sacrificed it is often preferable to operate at high pressures (and temperatures) to obtain sufficient solubility to make a process viable. The conditions cited for a specific process are often arrived at from an optimization of these opposing effects of selectivity and solubility. However, the selectivity that is exploited in extraction processes is sometimes an intrinsic property of the NCF solvent and is not always dramatically changed by the condition (e.g. in the selective extraction of triglycerides from phospholipids).

5.4.3.2 General principles

Effect of molecular structure

Any pragmatic assessment of a solvent extraction process must examine what type of molecules are soluble and to what extent. With NCFs the molecular structure of the solute is of major importance as small changes in molecular weight and functional groups can affect solubility to a greater extent than with liquid solvents. In fact the viability of many simple separation processes using NCFs can be realized without recourse to extensive solubility data covering a wide range of conditions. Some of the broad principles emerging from latest research work, carried out to correlate the solubilities of different classes of chemical compounds with molecular structure, are given below:

(1) Solubility is reduced by increasing polarity. A good illustration of this is to be found in the relative solubilities of ethanol and ethylene glycol in liquid CO_2. Whereas the former is completely miscible (M), increasing

the overall polarity by introducing a second hydroxyl group reduces the solubility to 0.2%. Miscibility with liquid CO_2 can be recovered by methylation of the OH groups, which reduces the polarity of the molecule.

(2) Solubility declines with increasing molecular weight and for any homologous series the solubility decreases rapidly beyond a given carbon number. Observations show that, the more polar alcohol has a lower carbon number 'cut off' than the parent alkane.

(3) Branching increases solubility. Thus 2, 6, 10, 14 tetramethyl pentadecane is miscible with liquid CO_2, whereas *n*-nonadecane is less than 3% (soluble).

(4) Unsaturation increases solubility as illustrated by 1-octadecene, which is about three times more soluble than its saturated homologue, *n*-octadecene.

(5) Aromaticity decreases solubility. This is well demonstrated by the progressive decrease in solubility in the series decalin-tetralin-naphthalene as aromaticity is introduced into the molecule.

Effect of temperature and pressure

For liquid solvents with low compressibilities the pressure has very little influence on solubility. A simple explanation of the effect of pressure on solubility in NCFs can be made in terms of the number of solute-solvent interactions which depends upon the density of the solvent medium. The overall shapes of solubility isotherms therefore often closely resemble density isotherms of the pure solvent. At very high pressures, restraints of packing can adversely perturb the preferred molecular orientations required for optimum solvation, and solubilities can then begin to decrease with increasing pressure.

As heats of solution are more often positive it is generally observed that solubilities in liquid solvents increase with temperature at constant pressure. However, with NCFs the situation is more complex since both density and temperature must be considered. A general statement governing the influence of these parameters is that 'the solubility increases with increasing temperature at constant density'. This generality is more universally obeyed than the alternative statement in terms of temperature alone.

5.4.4 Factors Determining the Efficiency of Ncf Extraction

The efficiency of a solvent extraction process can be judged by both the rate at which the process operates and its energy consumption. The rate limiting step in the process is most frequently the rate of mass transfer at the extraction stage, which in the case of leaching can be identified as the rate of dissolution of solute from a bed of particles.

The energy input is determined by:

(1) the concentration of solute in the NCF solvent leaving the extraction vessel (which determines the number of cycles required to complete the extraction);

(2) the differential conditions of pressure and temperature between the extraction and separation vessels (which determine the energy consumption per cycle).

In examining process efficiency, the extraction and separation stages will be treated separately.

5.4.4.1 Extraction stage

To illustrate the differences between NCF and liquids as extraction media the example of leaching of porous particles will be used. This involves the dissolution of a solute, initially evenly distributed throughout the porous matrix within the particles that are themselves insoluble. There are widespread examples of leaching in the food industry, including extraction of seed oils, decaffeination of coffee and isolation of flavor components from plant materials. The rate of removal of solute will depend upon many factors which are common to any solvent, including:

(1) the amount of solute in the particle;

(2) its distribution within the matrix;

(3) the particle size and shape;

(4) the geometry of the porous network.

To simplify the analysis it will be assumed that the solute to be removed is evenly distributed throughout a particle of regular shape (slab) as shown schematically in Fig. 5.5. Similar arguments apply for other geometries (e.g. sphere, cylinder).

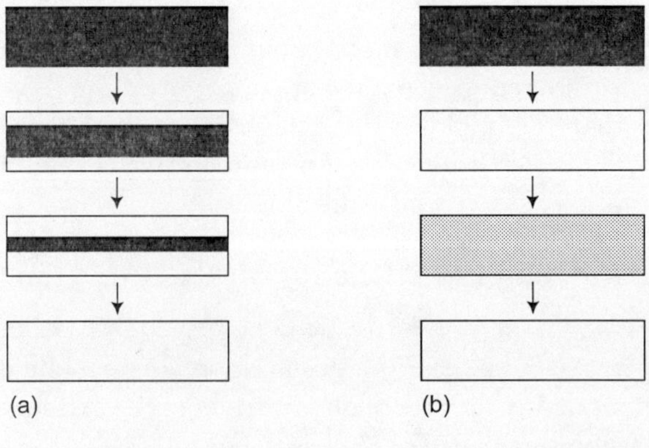

Fig. 5.5 Schematic representation of the extraction of solute from a slab by (a) the 'shrinking core' model and (b) the 'free diffusion' model

5.4.4.1.1 Mechanism of extraction

The solubility of the material to be extracted is clearly a prime factor in determining the effectiveness of a solvent in an extraction process. However, when solubilities are low it can also determine the mechanism by which the extraction proceeds. Because solubilities in SCFs are often low, conditions are frequently incurred which serve to establish a different mechanism (shrinking core) to that commonly found in liquid solvent extraction (free diffusion).

5.4.4.1.2 Solubility

The most important feature of the 'shrinking core' model is that the rate of extraction is determined not only by the diffusion coefficient but also by the solubility of the solute. Clearly when the solubility is low only a small portion of the solute can dissolve and participate in the diffusion process. Rates of extraction can therefore be very low when these conditions apply and this feature has been successfully exploited in the development of devices for controlled drug delivery. Generally solute loadings in SCFs are much lower than in liquid solvents, for which it is possible to choose from a wide range of solvent polarity in order to optimise solubility. The 'shrinking core' mechanism is therefore more likely to prevail in NCF solvents and is often responsible for the slow extraction rates associated with NCF leaching processes.

5.4.4.1.3 Diffusion coefficient

It is frequently reported that the higher diffusion coefficients in SCFs give rise to a more rapid extraction than could be achieved using liquid solvents. This is most certainly the case when the 'free diffusion' model applies. Unfortunately the low solubility in SCFs often serves to switch the mechanism to a 'shrinking core' model in which the enhanced diffusion in the SCF is offset by the diminished solubility. Therefore, although having other advantages, SCF extraction can sometimes be very slow in comparison with extraction processes using liquid solvents.

5.4.4.1.4 Adsorption

The extent by which a solute molecule in solution partitions to a surface is determined by the differential energy of interaction with the surface (adsorption energy) and solvent (solvation energy). Because solvation energies in SCFs are generally lower than in liquid solvents, partition coefficients to surfaces are often enhanced and adsorption effects are more pronounced. This has significant implications for extraction since adsorption serves both to retard the important rate-determining diffusion processes in leaching and also to reduce the maximum equilibrium concentration of solute that can be attained in the extraction vessel.

In the leaching process the simplest case of rapid adsorption/desorption of solute onto the internal pore surface can be accounted for by a linear isotherm.

In this model the diffusion coefficient is replaced by a modified diffusion coefficient (D^{eff}) inversely related to the adsorption coefficient (K_{ads}):

$$D^{eff} = D/(K_{ads} + 1)$$

Where;

K_{ads} = (concentration of adsorbed solute)/(concentration of freely diffusing solute)

When $K_{ads} > 1$ this will clearly reduce the rate of extraction.

In a fixed-bed extraction process, in the absence of adsorption, the maximum attainable concentration of solute in the solvent stream (C^{max}) is given by the solubility. This determines the minimum amount of solvent required to complete the extraction. In the presence of selective partitioning to the particle bed C^{max} is similarly attenuated:

$$C^{max} = S/(K_{abs} + 1)$$

The implications for extraction efficiency are twofold:

(1) More solvent will be required to complete the extraction, i.e. the process must operate through more cycles.

(2) Because the level of the solute in the SCF solvent is lower, a greater pressure differential is required to affect its precipitation in the separation stage of the process.

Both of these factors serve to increase the energy consumption of the process.

5.4.4.2 Separation stage

Since the solubility of all solutes in SCFs declines with decreasing pressure, the obvious way to separate the SCF at the separation stage is by decompression. This can, however, make heavy demands on energy input to the process, since the SCF then has to be recompressed on each cycle through the system. One solution to this problem has been to preferentially partition the solute to a coexisting solid or liquid phase in the separation vessel maintained at the same pressure as the extractor. This technique is particularly useful for processes in which the extracted components are undesirable contaminants present at low levels. Active carbon can then be employed to selectively adsorb the dissolved components from the CO_2 stream in an 'isobaric' process. In processes for the decaffeination of coffee, caffeine is often removed in this way by adsorption onto active carbon or by partitioning to a coexisting aqueous phase. Unfortunately most low molecular weight organic solvents are miscible with CO_2, and cannot be used for selective partitioning.

A feature of the separation stage, which is currently receiving much attention, is crystallisation. It is generally observed that the rapid decompression accompanying the separation of crystalline solids produces smaller crystals. This has formed the basis of a process for commutation of materials that would otherwise be difficult to handle by conventional milling techniques. Unlike temperature, pressure variation offers a rapid, isotropic response which can be used to control crystal size and morphology.

5.4.5 Equipment and Experimental Techniques Used in NCF Extraction

Although there is some degree of overlap, distinction will be made between equipment that has been designed for use in either fractionation or extraction processes. A description of large scale commercial-scale plant is also omitted since accurate information of plant operating in the commercial sector is often difficult to obtain and highly process specific. Relevant design features of large-scale SCF extraction equipment are available in specialist reviews, conference proceedings and directly from manufacturers.

Pilot plants with recirculation

A typical pilot plant configuration is shown schematically in Fig. 5.6. Although this type of equipment is designed for testing applications on a relatively small scale (< 10 kg) it includes most of the important features of a commercial-scale plant. The essential components of the plant are an extraction vessel, separation vessel, condenser and pump. The CO_2 is stored at its vapor pressure, in the condenser, as an NCL, and is pumped into the extraction vessel by a high-pressure liquid pump. The state of the CO_2 in the extraction/separation stages (SCF/NCL or gas) is determined by the temperature and pressure of the vessels. The temperature is controlled by a thermostatted re-circulating fluid flowing through 'heating jackets' surrounding the vessels and the pressure is maintained by pressure relief valves. Conditions of temperature, pressure and flow rate are continuously monitored and logged at strategic points throughout the system.

In an extraction process the material to be extracted is placed in the extraction vessel, which is purged with gaseous CO_2 to remove all air from the system. The extraction is then started by pumping the liquid CO_2 through a heat exchanger into the extraction vessel. The flow rate, which is determined by the pump, is usually set to allow sufficient residence time in the extraction vessel for equilibrium solubility to be attained. The solution then passes to the separation vessel, where conditions are set to minimize the solubility of the extracted components. This often involves decompression to a low density SCF state ($T>T_c$) or to a state on or below the liquid-gas coexistence line ($T<T_c$). Gaseous CO_2 then passes to the cooled condenser, where it is condensed and stored as a liquid.

Because pressure vessels incorporate thick metal walls, visual monitoring of the extraction process presents some problems. This is particularly pronounced

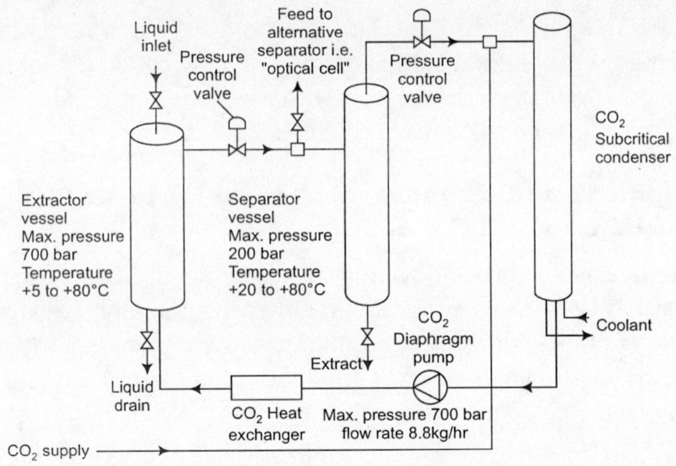

Fig. 5.6 Schematic diagram of an NCF CO_2 pilot plant capacity

at the separation stage, where it is necessary to monitor the amount and quality of extract obtained during the extraction. If the extract is a liquid it can be periodically removed from the separation vessel by a drain valve, but if a solid it is necessary to halt the extraction to remove the deposits by washing. Non-intrusive monitoring of the extraction can be achieved by spectroscopy (UV/Vis/NIR), if high pressure windows are incorporated in the flow line between the extraction and separation vessels. This technique is particularly useful for determining levels of extracted components with chromophoric groups, but is more amenable to optimizing established processes where the composition of the extract is known. It is also possible to switch a small sample of the extract stream into an analysis loop for chromatographic identification (GC, HPLC).

Provision of an additional high-pressure pump for dosing the CO_2 stream with entrainers is a common feature of most pilot plants. In case of liquid extraction the same pump can be configured to inject the liquid into the extraction vessel in a counter-current mode to establish a continuous process.

Pilot plants operating on this scale (2-10 liters) are useful when it is necessary to extract moderate quantities of material. However, for establishing the feasibility of a process it is often quicker and easier to use smaller-scale equipment.

5.4.6 Applications

The last decade has seen a rapid growth in research activity in the area of SCF extraction. Although many glowing reviews and hundreds of publications have appeared covering a wide variety of applications, there are surprisingly few processes and plant operating on a commercial scale. It is possible that many of the early publications overstressed the potential advantages of SCFs without addressing the limitations. It is certainly the case that many reports of applications do not differentiate between the extraction/identification of trace amounts of components and realistic quantities on which a process could be based. Without the necessary data it is often difficult or impossible to quantify the efficiency of a reported SCF extraction process in terms of mass transfer and throughput of CO_2. Against this background it is not surprising that many popular misconceptions abound regarding SCFs.

SCF extraction is an expensive process and, although offering many advantages (Table 5.2.), should not be used simply because it is a 'novel' technique. Unless use is being made of its unique features there is no rationale for its implementation if a cheaper separation process can meet the requirements of the separation as effectively. The one clear advantage that CO_2 does offer for food applications is its lack of toxicity. In the current climate of growing consumer concern regarding food safety this feature will undoubtedly promote its use in the food industry.

Table 5.2 Criteria determining implementation of SCF CO_2 extraction

Criteria	*Advantages*	*Disadvantages*
Economic	Can save energy Cheap Stable market	High capital cost.
Safety	Non-toxic* Non-flammable	High pressure.
Physical properties	Enhanced control through *T* and *P* (fractionation*). High vapor pressure enables separation at low *T**. Low viscosity can provide enhanced mass transfer.	Low solubilities.
Chemical properties	Non-oxidative. Environment*.	pH effects and hydrate formation when water present.

*= implication for improved quality.

The following review of applications includes examples of processes currently operating on a commercial scale and applications that could soon be implemented. Examples of applications showing novel features that serve to illustrate the potential of SCF extraction are also included.

Examples are:

1. Decaffeination of coffee and tea.
2. Seed oil extraction.

3. Purification of lecithin.
4. Lowering cholesterol levels in foods.
5. Fractionation of high-value oils and fats.
6. Extraction of flavors and fragrances.

5.5 LIQUID-LIQUID EXTRACTION

Liquid-liquid extraction as a technology has been used in the antibiotics industry for several decades and it is now beginning to be recognized as a potentially useful separation step in protein recovery and separation, particularly because it can readily be scaled-up and can, if necessary, be operated on a continuous basis. The physicochemical factors of the protein that determine partitioning are also starting to be understood. It is a reasonably high-capacity process and can offer good selectivity for the desired protein product. However, poor solubility of the large protein molecules in typical organic solvents restricts the range of solvents available for use in such a separation process.

5.5.1 Conceptual Theory

Many processes in chemical engineering require the separation of one or more of the components of a liquid mixture by treating the mixture with an immiscible solvent in which these components are preferentially soluble. In some cases purification of a liquid may be the function of the process, in others the extraction of a dissolved component for subsequent processes may be the important aspect. An example of the former is the preparation of the pure organic liquids from products of the oil industry. Liquid-liquid extractions may also be used as energy saving processes by, for example, eliminating distillation stages. It is possible, of course that the substance of interest may be heat-sensitive anyway and that distillation is accordingly an unacceptable process.

When separation by distillation is ineffective or very difficult, liquid extraction is one of the main alternatives to consider. Close-boiling mixtures or substances that cannot withstand the temperature of the distillation, even under a vacuum, may often be separated from impurities by extraction, which utilizes chemical differences instead of vapor pressure differences. For example, penicillin is recovered from fermentation broth by extraction with a solvent such as butyl acetate. Another example for liquid extraction is recovering acetic acid from dilute aqueous solutions; distillation would be possible in this case, but the extraction step considerably reduces the amount of water to be distilled.

When either distillation or extraction may be used, the choice is usually distillation, in spite of the fact that heating and cooling are needed. In extraction the solvent must be recovered for reuse (usually by distillation), and the combined operation is more complicated and often more expensive than ordinary distillation without extraction. However, extraction does offer more flexibility in choice of operating conditions, since the type and the amount of solvent can be varied as well as the operating temperature. In many problems, the choice between the methods should be based on a comparative study of both extraction and distillation.

In liquid-liquid extraction, as in gas absorption and distillation, two phases must be brought into contact to permit transfer of material and then be separated. Extraction equipment may be operated batchwise or continuous. The extract is the layer of solvent plus extracted solute and the raffinate is the layer from which solute has been removed. The extract may be lighter or heavier than the raffinate, and so the extract may be shown coming from top of the equipment in some cases and from the bottom in others. The operation may of course be repeated if more than one contact is required, but when the quantities involved are large and several contacts are needed, continuous flow becomes economical.

The rate at which a soluble component is transferred from one solvent to another will be dependent, among other things, on the area of the interface between the two immiscible liquids. Therefore it is very advantageous for this interface to be formed by droplets and films, the situation being analogous to that existing in packed distillation columns.

Two classes of solvents that appear to offer advantages for protein recovery for protein separations are aqueous polymer/salt (in some cases also polymer/polymer) systems and reverse micellar solutions. In both cases two phases are formed and the separation exploits the difference in partitioning of the proteins in the feed and extraction phases. In the aqueous polymer/salt separation systems the partitioning of the protein occurs between two immiscible aqueous phases; one rich in a polymer (usually polyethylene glycol, PEG) and the other in a salt (e.g. phosphate or sulphate). These systems show a non-denaturing solvent environment, small interfacial resistance to mass transfer, relatively high protein capacity and high selectivity. On the other hand reverse micelles exploit the solubilising properties of surfactants that can aggregate in organic solvents to form so-called inverted or reverse micelles. These aggregates consist of a polar core of water and the solubilised protein stabilised by a surfactant shell layer. For protein extraction, one phase is the aqueous feed solution, the other the reversed micellar phase that acts as the extractant. They have several of the advantages quoted for aqueous two-phase systems. Various liquid-liquid extractors abide a range of advantages and disadvantages with them which are displayed below in Table. 5.3.

Table 5.3 Advantages and disadvantages of the various liquid-liquid extractor types

Unit of Operation	*Advantages*	*Disadvantages*
Mixer-Settler	• Efficient • Low head room • Induces good contacting • Can handle any number of stages	• Large floor • High set-up costs • High operation costs
Columns (without agitation)	• Small investment costs • Low operating costs	• High head room • Difficult to scale up from lab • Less efficient than mixer-settler
Columns (with agitation)	• Good dispersion • Low investment costs • Can handle any number of stages	• Difficult to separate small density differences • Does not tolerate high flow ratios
Centrifugal Extractors	• Can separate small density differences • Short holding time • Small liquid inventory	• High set-up cost • High operating and maintenance costs • Cannot handle many stages

The suitability of using foam separation as well as gas aphrons as novel separation techniques for proteins are presently under investigation.

5.5.1 Aqueous Two-phase Separation

Partitioning in two aqueous phases can be used for the separation of proteins from cell debris as well as for purification from other proteins. Partitioning can be done in a single step or as a multistage process. Differences in partition coefficients, however, between the different proteins can be high; hence one step tends to be sufficient (usually one for extraction and one for elution or back-extraction). The use of affinity partitioning can greatly enhance the specificity of the extraction. A typical process for extraction of a protein into the top PEG phase in a first stage and the back extraction into a bottom salt phase (e.g. phosphate or sulphate) in a second 'back extraction' step from a cell homogenate that includes recycle of the PEG phase is shown in Fig. 5.7.

Most soluble and particulate material partitions to the lower, more polar (e.g. salt) phase and the protein of interest partitions to the top less polar phase, usually PEG. Separation of actual proteins in such systems is based on manipulating the partition coefficient (*K*) by altering parameters such as average molecular weight of the polymer, type of phase forming salt used for the heavy phase, the types of ions included in the system and ionic strength of added salts (e.g. NaCI). Fig. 5.8. shows that the partition coefficient of α-amylase is a strong function of the presence of NaCl in a PEG/sulphate system.

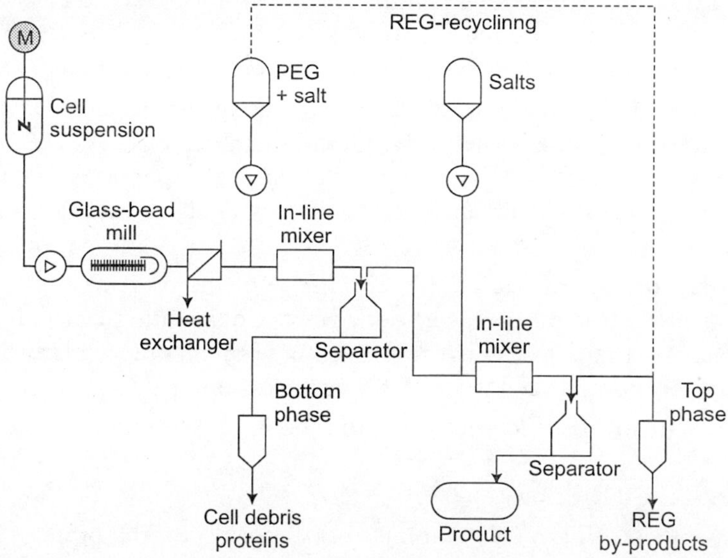

Fig. 5.7 Scheme of enzyme purification by liquid-liquid extraction. The cells are disrupted by wet milling, and after passing through a heat exchanger, PEG and salts are added into the process stream of broken cells. After mixing and obtaining of equilibrium the phase system is separated, the outflowing bottom phase is going to waste. The product-containing PEG-rich top phase goes to a second mixer after addition of more salt to the process stream. The product is recovered in the resulting bottom phase while the concentrated PEG solution (upper phase) goes to waste or is recycled.

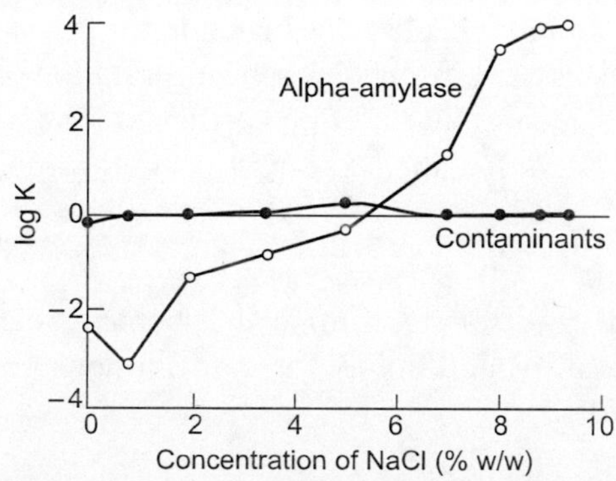

Fig. 5.8 Partition behaviour of α-amylase (log $K_á$)and contaminant protein (log K_c) from industrial supernatant from *E. subtilis* fermentation in PEG 4000/Sulphate systems as a function of added NaCl concentration at pH 7 and a phase volume ratio of 1.

For extraction of the α-amylase from its contaminants a high concentration of NaCl is used in the first extraction stage, whereas a low concentration of NaCl in the back-extraction stage will allow the recovery of á-amylase into the bottom sulphate phase as shown in Fig. 5.7.

The partition coefficient *(K)* is defined as the concentration of a particular protein in the lighter phase divided by the concentration in the heavier phase. The main factors that determine partition depend on the type of system used:

(1) **Hydrophobicity:** Differences in the surface hydrophobicity between proteins are exploited when partitioning them in PEG/salt two-phase systems. Typical systems that exploit a protein's hydrophobicity are PEG/phosphate and PEG/sulphate with addition of a high concentration of NaCl (e.g. 10%).

(2) **Size-dependent partition:** Molecular size of the proteins or surface area of the particles to be partitioned is the dominating factor. It has been shown that for PEG/Dextran systems a protein's molecular weight is inversely proportional to its partition coefficient.

(3) **Electrochemical:** Electrical potential between the phases is used *to* separate molecules or particles according *to* their charge. This is demonstrated in below.

(4) **Biospecific Affinity:** The affinity between sites on the proteins and ligands attached to one of the phase polymers is used for separation. Dyes, inhibitors, fatty acids, glutathione, Protein A and several other ligands have been used. Particularly impressive results for selective partitioning have been obtained with metal ions both in PEG/ Dextran but also in PEG/ salt systems.

(5) **Solubility Dependent:** In addition it is important to know the actual concentration of the protein in the extractant phase. In typical PEG/ salt systems, protein solubility tends to be higher in the PEG and lower in the salt phase. It is clear that, in the region near saturation of the protein in one of the phases, a constant partition coefficient is no longer observed.

According to this, it is possible to split the partition coefficient into different terms:

$$K = K_{hfob} K_{el} K_{mw} K_{aff} K_{sol}$$

where hfob, el, mw, aff, and sol stand for hydrophobicity, electrostatic, size (molecular weight), affinity and solubility contributions to the partition coefficient.

5.5.2 Liquid Membrane Separations

Liquid membrane extraction is a relatively new separation technology which has significant potential for the selective separation and concentration of low molecular weight chemicals produced by fermentation and used in the food-processing industries. Separation is achieved by the transport of the solute from a feed phase across a film of organic solvent into a stripping phase. Examples of products successfully extracted using liquid membranes include organic acids such as citric acid, lactic acid, and amino acids such as L-phenylalanine.

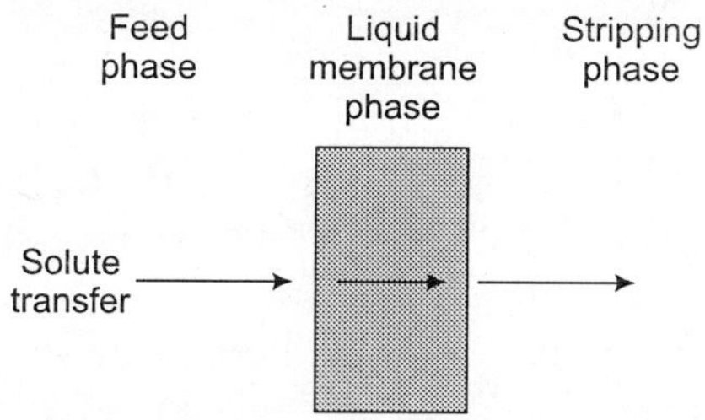

Fig. 5.9 Schematic diagram of liquid membrane processes.

Currently, the recovery and purification processes for these species involve several steps specific to individual manufacturers. There is, therefore, scope for the application of liquid membranes which are generally one-step processes, and can simultaneously separate and concentrate the solute. A schematic diagram of a liquid membrane process is shown in Fig. 5.9.

The liquid membrane consists of the organic solvent which separates the two aqueous phases (the feed and stripping phases), and contains a carrier species to enhance both selectivity and rates of extraction. Aliphatic diluents are generally preferred as the solvent because of their lower solubility in water. In an ideal situation the solvent should have no solubility in water to ensure that there is no aqueous phase contamination by trace organics. There are two main configurations by which liquid membrane extraction can be exploited, as discussed in the next two paragraphs.

A supported liquid membrane (SLM, Fig. 5.10.) can be achieved by impregnating a porous solid film with an organic solvent, which is held in place by capillary forces that exist within the pores. In order for the membrane pores to be effectively wetted; the surface tension of the solvent must be less than the critical surface tension of the membrane polymer. The membrane separates an aqueous phase, initially containing the required species, from an aqueous phase into which the solute is extracted, the stripping phase. The solid supports used are generally microporous polymeric films, e.g. polypropylene, polysulphone, or other hydrophobic materials. Typical dimensions are a membrane thickness of 25-50 μm, with pore sizes of 0.02-1.0 μm.

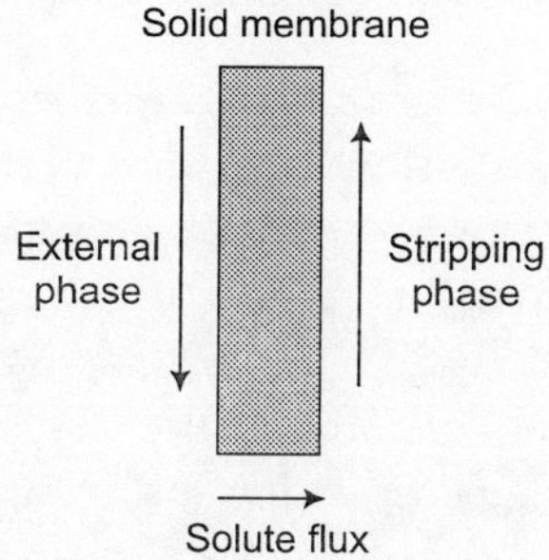

Fig. 5.10 Schematic diagram of supported liquid membrane (SLM)

An emulsion liquid membrane (ELM, Fig. 5.11.) is formed by creating, under high shear, a dispersion of the stripping phase within the organic solvent which forms a nonporous film around the stripping phase droplets. The emulsion thus formed (stabilised by a surfactant) is dispersed into the feed phase containing the solute, which is then transported into the stripping phase. Depending on the dispersion conditions the globule diameter is 1-2 mm and the internal phase droplets are micron sized. The two aqueous phases cannot physically contact each other and the solute is transported into the internal phase droplets by diffusion through the stabilised solvent film. The use of a chemical reagent in the stripping phase, which reacts with the extracted solute, prevents the solute from diffusing back across the membrane phase. This strategy allows the removal of virtually all of the solute from the feed solution, which makes emulsion liquid membrane extraction very attractive for the recovery of solutes formed in low concentration. After the extraction step the solute is recovered by allowing the emulsion and feed phases to separate, by settling under gravity, then removing the emulsion and breaking it to release the separated solute from the membrane phase components. Electrostatic splitting is generally used for de-emulsification as the membrane components can be recycled for further use.

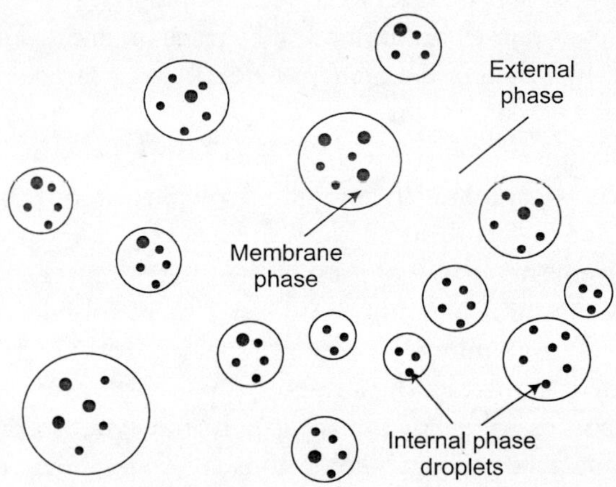

Fig. 5.11 Schematic diagram of emulsion liquid membrane (ELM)

ELM systems give rise to very fast extraction kinetics and allow the use of conventional liquid-liquid extraction equipment; they are also prone to emulsion swelling, which gives rise to dilution and instability problems, The necessity to make and break an emulsion does not arise with the SLM system; however, this configuration has slower kinetics, and loss of the membrane phase may occur. In summary, liquid membrane processes offer high separation factors, low capital and operating costs, a lower solvent inventory than solvent extraction, ease of scale-up and the possibility of continuous operation.

The key to selectivity in liquid membrane extraction is the use of a carrier species incorporated in the organic solvent to increase the solute solubility: by introducing a 'carrier' molecule into the membrane phase, the solute solubility is increased by the reversible formation of a membrane-soluble carrier-solute complex. This results in faster mass transfer rates, and selectivity is introduced into the extraction as the carrier-solute reaction can be selective. This is known as facilitated transport. The use of a carrier enhances selectivity by the formation of a reversible complex between the carrier and the solute, which is only soluble in the organic solvent. This is particularly effective for the recovery of charged solutes which may be poorly soluble in the organic solvent. Many of the carriers so far employed in liquid membrane processes are extractants used in conventional liquid-liquid extraction, e.g. secondary and tertiary amines, and phosphorus-containing extractants.

5.6 SOLID-LIQUID EXTRACTION

5.6.1 Theory

Extraction is a term used for an operation in which a constituent of a liquid is transferred to another liquid (solvent). The term solid-liquid extraction is restricted to those situations in which a solid phase is present and includes those operations frequently referred to as leaching, lixiviation and washing.

Extraction always involves two steps:

1. Contact of the solvent with the solid to be treated so as to transfer the soluble constituent (solute) to the solvent;
2. Separation or washing of the solution from the residual solid.

These two steps may be conducted in separate equipment or in the same piece of equipment.

Liquid always adheres to the solid which must be washed to prevent either the loss of solution if the soluble constituent is the desired material or the contamination loss of the solids if these are the desired material. The complete process also includes the separate recovery of the solute and solvent. This is done by another operation such as evaporation or distillation.

The equipment used for solid-liquid extraction is classified according to the manner in which the first step is done. The term *solid bed* or *fixed bed* refers to any operation in which the solid particles are kept in relatively fixed positions with respect to each other while the solvent flows through the bed of solid particles, whether or not the bed of solid material remains stationary with respect to the earth during the extraction. The opposite of this situation, called *dispersed contact*, refers to any operation in which the solid particles, suspended in the fluid, are in motion relative to each other and to the solvent during the time of contact.

5.6.2 Leaching

Solid-liquid extraction (leaching) is the process of removing a solute or solutes from a solid by using of liquid solvent. Leaching is widely used in chemical industries where mechanical and thermal methods of separations are not possible or practical. Extraction of sugar from sugar beets, oil from oil bearing seeds, production of a concentrated solution of a valuable solid material are typical industrial examples of leaching. Leaching process can be considered in three parts:

1. Diffusion of the solvent through the pores of the solid.
2. Solute dissolves in the solvent.
3. Transfer of the solution from porous solid to the main bulk of the solution.

In fixed-bed contacting the solid particles are stationary in a tank, in which the solvent is allowed to percolate through the bed of undissolved solids. In leaching by the soxhlet apparatus, multiple contacts of solids with the fresh solvent are performed at each stage of operation. This is equivalent to multiple cocurrent system, as described in the following diagram.

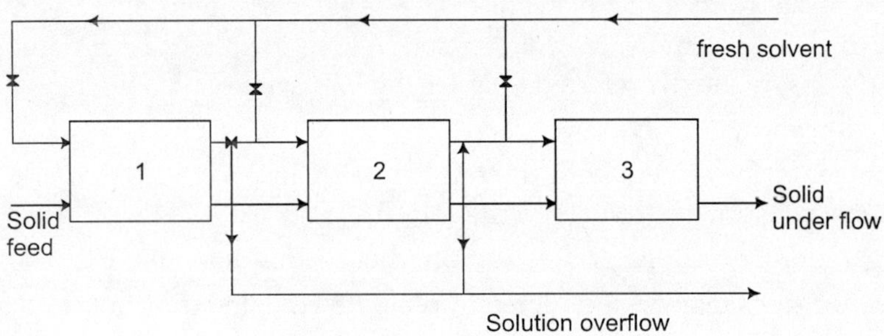

Fig. 5.12 Multiple cocurrent system

5.7 SOLID BASED SEPARATIONS: ADSORPTION

5.7.1 Introduction

Adsorption is a process that occurs when a gas or liquid solute accumulates on the surface of a solid or a liquid (adsorbent), forming a molecular or atomic film (the adsorbate). Adsorption, ion exchange and chromatography are sorption processes in which certain adsorptives are selectively transferred from the fluid phase to the surface of insoluble, rigid particles suspended in a vessel or packed in a column. Similar to surface tension, adsorption is a consequence of surface energy. In a bulk material, all the bonding requirements (be they ionic, covalent or metallic) of the constituent atoms of the material are filled. But atoms on the (clean) surface experience a bond deficiency, because they are not wholly surrounded by other atoms. Thus it is energetically favorable for them to bond with whatever happens to be available. The exact nature of the bonding depends on the details of the species involved, but the adsorbed material is generally classified as exhibiting physisorption or chemisorption.

Adsorption is operative in most natural physical, biological, and chemical systems, and is widely used in industrial applications such as activated charcoal, synthetic resins and water purification. The most common industrial adsorbents are activated carbon, silica gel, and alumina, because they present enormous surface areas per unit weight. Activated carbon is produced by roasting organic material to decompose it to granules of carbon - coconut shell, wood, and bone are common sources. Silica gel is a matrix of hydrated silicon dioxide. Alumina is mined or precipitated aluminum oxide and hydroxide. Although activated carbon is a magnificent material for adsorption, its black color persists and adds a grey tinge if even trace amounts are left after treatment; however filter materials with fine pores remove carbon quite well. Adsorption is different from absorption, in which a substance diffuses into a liquid or solid to form a solution. The term sorption encompasses both processes, while desorption is the reverse process.

5.7.2 Principle of Adsorption

Adsorption is described as the enrichment of one or more components in the interfacial layer, i.e. an excess of molecules exists at the adsorbate/adsorbent interface, upon exposure of an adsorbing solid to a gas or vapor. It is the selective collection and concentration onto solid surfaces of certain molecules contained in a vapor or gas stream. Hence, vapors or gases referred to as adsorbates when adsorbed, even of mixed systems and at low concentrations, may be captured, often selectively, and removed from the effluent stream using a material of the category of adsorbents.

Adsorption is usually described through isotherms, that is, functions which connect the amount of adsorbate on the adsorbent, with its pressure (if gas) or concentration (if liquid).

The first isotherm is due to Freundlich and Küster (1894) and it is a purely empirical formula valid for gaseous adsorbates:

$$\frac{x}{m} = kP^{\frac{1}{n}}$$

Where, x is the adsorbed quantity, m is the mass of adsorbent, P is the pressure of adsorbate and k and n are empirical constants for each adsorbent-adsorbate pair at each temperature. The function has an asymptotic maximum. As the temperature increases, the adsorbed quantity rises more slowly and more pressure is required to achieve the maximum.

5.7.3 Adsorption Isotherms

Langmuir isotherm

In 1916, Irving Langmuir published a new isotherm for gases adsorbed on solids, which retained his name. It is an empirical isotherm derived from a proposed kinetic mechanism. It is based on four hypotheses:

1. The surface of the adsorbent is uniform, that is, all the adsorption sites are equal.
2. Adsorbed molecules do not interact.
3. All adsorption occurs through the same mechanism.
4. At the maximum adsorption, only a monolayer is formed: molecules of adsorbate do not deposit on other, already adsorbed, molecules of adsorbate, only on the free surface of the adsorbent.

These four points are seldom true: there are always imperfections on the surface, adsorbed molecules are not necessarily inert, the mechanism is clearly not the same for the very first molecules as for the last to adsorb. The fourth condition is the most troublesome, as often more molecules can adsorb on the monolayer, but this problem is solved by the BET isotherm.

Frumkin isotherm

Frumkin isotherm is an extension of Langmuir isotherm. It states that adsorbed molecules do interact and affect further adsorption by either repulsion or attraction of molecules.

BET isotherm

Often molecules do form multilayers, some are adsorbed on already adsorbed molecules and the Langmuir isotherm is not valid. In 1938 Stephan Brunauer, Paul Emmett and Edward Teller developed an isotherm that takes into account that possibility. The biggest step in BET isotherm is to consider that the successive equilibria for all the layers except for the first are equal to the liquefaction of the adsorbate. Langmuir isotherm is usually better for chemisorption and BET isotherm works better for physisorption.

5.7.4 Adsorption Systems: Expanded Bed Adsorption

Expanded bed adsorption or fluidised bed adsorption is a new technique that in one step accomplishes removal of whole cells and cell debris, concentration and initial purification of the target protein. It is based on the design of an adsorbent whose density enables the formation of a stable expanded or fluidised bed (two- or threefold expansion) that allows the establishment of a plug-flow concentration profile across the column. Binding capacities for proteins are similar to those obtained in fixed bed adsorption and linear velocities can be higher than 300 cm/h. Pharmacia has recently started to commercialize expanded bed adsorption under the trade name streamline. This operation allows the fusing of two or three operations, namely clarification, concentration and capture, into one operation. At the outset, the adsorbent is sedimented in the column. The bed is then expanded by pumping buffer upward through the column. Once the bed is expanded, the buffer is replaced with crude feed which contains the products, cells, cell debris, contaminants and other particulates. After the product is adsorbed, loosely bound material is flushed with a buffer wash and the direction of the liquid flow is reversed with the change to elution buffer and the product is eluted as in packed bed chromatography. If the product is extracellular the process stream is the crude fermentation broth and if it is intracellular the stream is the crude homogenate.

5.7.5 Continuous Adsorption Recycle Extraction

Continuous adsorption recycle extraction was recently adapted for biotechnological applications as a downstream process for efficient separation of proteins from crude feedstocks to be carried out continuously. Modem biotechnology involves relatively small-scale processes that favor batch procedures, usually following the concept that a batch is well defined (from beginning to end) and that this will pose fewer problems with regulatory approval for therapeutic use. However, it is well known that in the chemical and biochemical process industries batch-to-batch variations can be substantial. It is also well known that a continuous process can operate under steady-state conditions minimizing batch-to-batch variations. In addition, a continuous process is much more appropriate for implementing a control strategy in order to maintain constant conditions more readily.

Continuous adsorption recycle extraction, also called CARE, is based on the continuous protein adsorption to solid phase supports and features two well-mixed reactors with solids recycle, as shown in Fig. 5.13. The adsorbing stage takes place in the first reactor (adsorber) where the liquid feed is contacted with the adsorbent; desorption of the protein to be purified is obtained in the second reactor (desorber) by maintaining appropriate

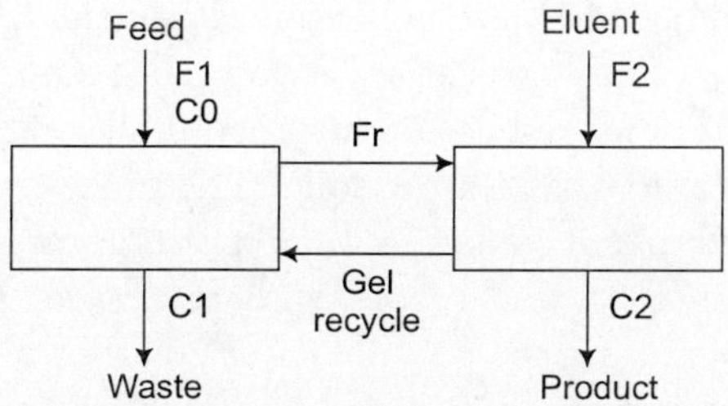

Fig. 5.13 Continuous adsorption recycle extraction (CARE) process.

conditions and by the action of a suitable eluting solvent. The adsorbent beads are recycled to the adsorber reactor while the product is continuously removed. The system's performance is determined by the nature of the adsorbent and the feed material, flow rates and reactor volume and the conditions of the adsorption and de sorption stages.

Where, C0 = Feed protein concentration; Fr = Gel recycle rate; F1 = Feed flow rate; F2 = Eluent flow rate; C1 = Waste stream concentration; C2 = Product stream concentration.

During operation, the feed composition and flow rate may experience uncontrolled variations which may upset the steady state. This is also true for adsorbent binding capacity and even recycle flow rates and composition. Hence, appropriate control schemes are needed to meet product specifications and achieve operational stability in the face of potential external and internal disturbances.

5.7.6 Types of Adsorption

Adsorption is divided into the two sub-categories of physical adsorption (physisorption) or van der Waals adsorption and chemical adsorption (chemisorption) and the adsorption process can be determined whether chemical bonds are formed during the process. Physisorption is applicable to all adsorbate-adsorbent systems provided the conditions of pressure and temperature are suitable whereas chemisorption may only occur if the system is capable of making a chemical bond.

5.7.6.1 Physical adsorption (physisorption)

The process is a dynamic one where an equilibrium state exists with molecules and the interaction between the adsorbate and adsorbent. No chemical bonds are formed during physical adsorption; attraction between the adsorbate and adsorbent exists by the formation of intermolecular electrostatic, such as London dispersion forces, or van der Waals forces from induced dipole-dipole interactions, or may be dependent on the physical configuration of the adsorbent such as the porosity of activated carbons. Dispersion forces are the result of rapid fluctuations in the electronic density of one adsorbent molecule inducing an electrical moment in a second atom. If the adsorbate possesses a permanent dipole, or even a multipole, then additional interactions may occur, as charge distributions are induced in the adsorbent and interactions of these moments with any permanent field of the solid. The process is a very general one and is analogous with that of condensation. Physisorption occurs to varying extents for all adsorbates, gases and vapors, with all adsorbing solids and the effect increases with decreasing temperature or increasing pressure. Physical adsorption is based on certain basic considerations and adsorption on a heterogeneous surface, that is a surface on which the sites are different, occurs at the sites of highest adsorption potential. The process of physical adsorption into the microporous structure of activated carbon follows the theory of Dubinin.

The mechanism of adsorption is dependent upon the size of the admolecule in comparison with the pore width due to the energetic interactions between the chosen adsorbate and the pores. Admolecules initially adsorb into the pores with the highest energy, ignoring activated diffusion effects, then adsorption proceeds via filling of progressively larger, or decreasing energy, porosity. Some pores are capable of accommodating two or three admolecules and, therefore, may undergo co-operative adsorption effects by reducing the volume element thus increasing the energy and adsorptive potential of the pore. The process of adsorption is always exothermic due to the increased ordering of the adsorbate on the adsorbent surface, reducing the entropy, as:

$$DG = DH - TDS$$

Thus the amount adsorbed should decrease with increasing temperature as a reduction in the thermal energy supplied to the process, by Le Chatelier's principle, favours the exothermic process of adsorption increasing the equilibrium uptake, except in the case of activated diffusion. It has been proposed by Lamond and Marsh, by the interpretation of data for physical adsorption of nitrogen on both polar and non-polar surfaces that physical adsorption is independent of the surface chemistry of the adsorbent.

5.7.6.2 Chemical adsorption (chemisorption)

Chemisorption involves the transfer of electrons between the adsorbent and the adsorbate with the formation of chemical bonds, by chemical reaction, between the two species causing adhesion of the adsorbate molecules. Chemical adsorption is far less common than physical adsorption and due to the chemical bonds formed regeneration of the adsorbent for subsequent re-use is often difficult or impossible. Due to the fact that chemical bonds are formed during the adsorption process, desorption of the adsorbed phase may yield products which are chemically different to the original adsorbate. For example oxygen may chemically bond to the surface of a carbon, which upon desorption may evolve CO and CO_2 as products.

5.7.7 Factors Influencing Adsorption

5.7.7.1 Surface area

The extent of adsorption is proportional to specific area. Specific area can be defined as that portion of the total area that is available for adsorption.

Table 5.4 Characteristics Associated with Physical/Chemical Adsorption

	Physical Adsorption	*Chemical Adsorption*
Heat of Adsorption / $kJmol^{-1}$	20-40 c.f. heats of liquefaction	> 80 c.f. bulk-phase chemical reactions
Rate of Adsorption (at 273K)	Fast	Slow
Temperature Dependence of Uptake (with Increasing T)	Decreases	Increases
Desorption	Easy- by reduced pressure or increased temperature	Difficult - high temperature required to break bonds
Desorbed Species	Adsorbate unchanged	May be different to original adsorptive
Specificity	Non-specific	Very Specific
Monolayer Coverage	Mono or multilayer condition dependent	Monolayer

5.7.7.2 Nature of the Adsorbate

(a) Adsorption of a solute is inversely proportional to its solubility in the solvent. The greater the solubility, the stronger the solute-solvent bond and the small the extent of adsorption. In general, the solubility of any organic compound in water decreases with increasing chain length, because the compound becomes more hydrocarbon-like as the number of carbon atoms becomes greater. Adsorption of aliphatic series of organic acids increases in the order formic-acetic-propionic-butyric, whereas the order is reversed for adsorption from toluene.

(b) Molecular sizes. If the rate is controlled by intraparticle transport, the reaction generally will proceed more rapidly the smaller the adsorbate molecule, within a given class of compound.

(c) Charge (Ionic species). As long as the compounds are structurally simple. adsorption is at minimum for the charged species and at a maximum for the neutral species. As compounds become more complex, the effect of ionization becomes of decreasing important. Decreasing adsorption with increasing ionization has been observed for many organic acids. Adsorption of propionic acid on carbon decreases markedly with increasing pH to a minimum in the range pH 3.5 to 5.5 (succinic acid to a min. in the range pH 4.0 and 7.0, caproic acid between pH 4.8 and 7.5). Studies on amphoteric compound indicate an adsorption maximum at the isoelectric point, or that pH at which both the acidic end and the basic end of the compound are ionized and the compound bears a net charge of zero.

(d) pH. Because hydrogen and hydroxide ions are adsorbed quite strongly, the adsorption of other ions is influence by the pH of the solution. In general, adsorption of typical organic pollutant from water is increased with decreasing pH.

(e) Temperature. Adsorption reaction are normally exothermic, thus the extent of adsorption generally increases with decreasing temperature.

(f) Adsorption of Mixed Solutes. The degree of mutual inhibition of competing adsorbates should be related to the relative sizes of the molecules being adsorbed, to the relative adsorptive affinities, and to the relative concentrations of solutes.

(g) Nature of the Adsorbent. The physicochemical nature of the adsorbent can have profound effects on both rate and capacity for adsorption. Commercial carbons can be prepared from a variety of raw materials, including wood, lignite, coal, petroleum residues, and nut shells. The raw material is generally activated in an atmosphere of CO_2, CO, O_2, H_2O, vapor, air, or other selected gases, at a temperature between 300 and 100°C often followed by quenching in air or water. The most characteristic physical property of activated carbon is its extremely large surface area, which is composed mainly or surfaces bordering inner pore spaces.

Different types of industrial adsorbents generally fall into three classes:

- Oxygen-containing compounds – Are hydrophilic and polar, including materials such as silica gel and zeolites.
- Carbon-based compounds – Are hydrophobic and non-polar, including materials such as activated carbon.
- Polymer-based compounds – Are polar or non-polar functional groups in a porous polymer matrix.

The adsorbents are used usually in the form of spherical pellets, rods, moldings or monoliths with hydrodynamic diameter between 0.5 and 10 mm. They must have high abrasion resistance, high thermal stability and small micropore diameter, which results in higher exposed surface area and hence high capacity of adsorption. The adsorbents must also have a distinct macropore structure which enables fast transport of the gaseous vapors.

6

CHAPTER

PRECIPITATION

6.1 INTRODUCTION

Precipitation is widely used in downstream processing of biological products, such as proteins. This unit operation serves to concentrate and fractionate the target product from various contaminants. For example, in the biotechnology industry protein precipitation is used to eliminate contaminants commonly contained in blood. Academic research on protein precipitation explores new protein precipitation methods.

Precipitation is a procedure where the addition of ionic solution to an ionic fermentation broth forms insoluble particles, where the desired product is usually contained within those particles. Ionic fermentation broth usually consists of enzymes or proteins. The ways to precipitate out a product can vary from single pH and temperature changes to chemical reactions involving metal ions. Precipitation reactions are carried out in reactors, continuous and batch.

The development of techniques and methods for protein purification has been an essential pre-requisite for many of the advancements made in biotechnology. Protein purification varies from simple one-step precipitation procedures to large scale validated production processes. Often more than one purification step is necessary to reach the desired purity. The key to successful and efficient protein purification and precipitation is to select the most appropriate techniques, optimize their performance to suit the requirements and combine them in a logical way to maximize yield and minimize the number of steps required. Most purification schemes involve some form of chromatography. As a result chromatography has become an essential tool in every laboratory where protein purification is needed. Different chromatography techniques with different selectivity's can form powerful combinations for the purification of any biomolecule. The development of recombinant DNA techniques has revolutionized the production of proteins in large quantities. Recombinant proteins are often produced in forms which facilitate their subsequent chromatographic purification.

6.2 REVIEW OF PRECIPITATION PROCESSES

Methods for protein separation may be divided broadly into those which divide the protein between two phases, usually but not always a solid (precipitate) and liquid (supernatant), and those which separate proteins by different rates of movement through some material, usually a chromatographic column but also including electrophoresis.

Enzymes and other proteins are highly charged molecules, and can be precipitated with appropriate charge neutralizing chemicals. Once their charges are broken, they form aggregates and settle down as precipitate.

When, an acid or base is added, the enzyme protein can be brought to its isoelectric pH. At this pH, there is no net charge on enzyme molecules and electrostatic repulsion between them is low so that they tend to aggregate. Therefore, adjusting the pH to the isoelectric point of a protein causes its precipitation.

Acids and bases, however, often inactivate the enzyme, so that their use for precipitation is not recommended in most cases. Instead ammonium sulphate and other salts are used for precipitation in a process called 'salting out'. Salts can change the structure of the solvent, which can lead to large changes in protein conformation by altering the electrostatic interactions between charged groups on the protein surface. The salt also competes with the protein for solvent molecules and thereby lowers its solvation. In large scale enzyme precipitation, use of many other neutral salts is preferred over ammonium sulphate, which is corrosive and releases NH3 at higher temperatures.

Some organic solvents like acetone, methanol and ethanol are also used for enzyme precipitation, since water miscible solvents decrease the solubility of proteins, leading to precipitation. They are cooled upto 40 60°C before their use, and precipitation is carried out at O°C, because precipitation at room temperature causes denaturation of the enzyme, in most cases. Organic solvents are added drop by drop to avoid local concentration.

Water soluble non ionic polymers such as polyethylene glycol, alginate, pectate, carboxymethyl cellulose, polyacrylic and polymeta acrylic acids, etc. also cause enzyme precipitation. Polyethyleneimine is also widely used as protein precipitant at large scale. They primarily act through the removal of solvent sphere of the enzyme protein.

6.3 PRINCIPLE AND THEORY

We should consider briefly the interactions which lead to precipitation. Proteins might stick to each other through one of three forces: electrostatic, hydrophobic, and van der Waals. Electrostatic forces operate at long range, but between like molecules are repulsive rather than attractive, since like molecules have the same charge and repel each other. Only when the distribution of surface charge on molecules is very non-random, positive in some areas of the surface, negative in others, is there net electrostatic attraction; and this may lead only to associations of two proteins, not to large scale aggregates which require interactions between many molecules. What we observe is that proteins are most likely to aggregate and precipitate at their isoelectric points, where they bear no net charge and do not repel each other. However, they can be precipitated with non-protein polyanions such as polyacrylate, or protamine sulfate for cationic proteins and ribonucleoproteins.

The main mode of general protein-protein interaction is hydrophobic. Nonpolar patches of the protein surface are shielded by water molecules arranged in an ordered structure; when two non-polar patches come together, the water molecules are expelled and go to a free, less ordered state, which increases their entropy. This increase in the entropy of water molecules, as the number of them solvating hydrophobic surfaces decreases, is the main driving force for protein association. This is all the more effective when the proteins are denatured.

Repulsive electrostatic forces form when proteins are suspended in an electrolyte solution. These repulsive forces between proteins prevent aggregation and facilitate dissolution. Solvent counter ions migrate towards charged surface residues on the protein, forming a rigid matrix of counter ions attached to the protein surface. The adjacent solvation layer, which is less rigid, consists of a decreasing concentration profile of the counter ions and an increasing concentration profile of the co-ions. In effect, the protein's potential to engage in ionic interactions with each other will decrease. Proteins will be less likely to form aggregates. Water molecules can have a similar effect. Water forms a solvation layer around hydrophilic surface residues of a protein. Water establishes a concentration gradient around the protein, with the highest concentration at the protein surface. This water network has a damping effect on the attractive forces between proteins.

Dispersive or attractive forces exist between proteins through permanent and induced dipoles. For example, basic residues on a protein can have electrostatic interactions with acidic residues on another protein. However, solvation by ions in an electrolytic solution or water will decrease protein-protein attractive forces. Protein accumulation and precipitation can be enhanced by decreasing the hydration layer around the protein. The purpose of the added reagents in protein precipitation is to reduce the hydration layer.

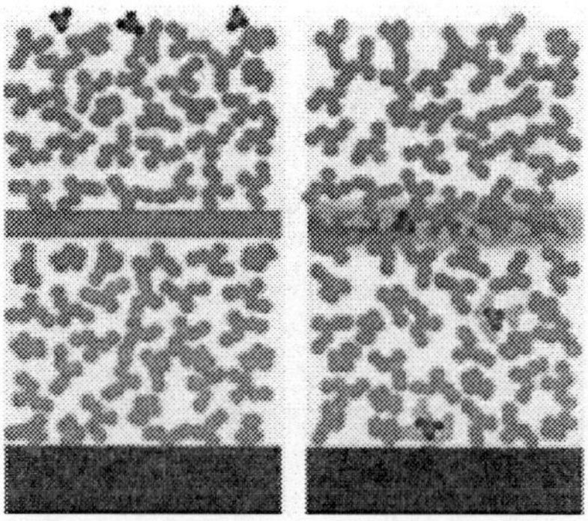

Fig. 1 Solvation of ions using oil water interface

Protein precipitation by removal of the shell of hydrating water, as in $(NH_4)_2SO_4$ or PEG precipitation is generally reversible, but lyophilization or even solvent precipitation may be irreversible. Typical agents used in precipitation are:

1. Acids and base to change pH of a solution till isoelectric point is reached where solubility of product decreases.
2. Salts such as ammonium sulphate removes water from the surface of compound thus hydrophobic regions come closer and aggregates
3. Organic solvent such as acetone changes dielectric properties of solution
4. Non-ionic polymers like polyethylene glycol works in same fashion as organic solvents

6.4 PRECIPITATE FORMATION

Protein precipitate formation occurs in a stepwise process. The addition of a precipitating agent and steady mixing destabilizes the protein solution. Mixing causes the precipitant and the target product to collide. Enough mixing time is required for molecules to diffuse across the fluid eddies. During the following nucleation phase, submicroscopic sized particles are generated. Growth of these particles is under Brownian diffusion control. Once the growing particles reach a critical size (0.1 μ to 10 μm for high and low shear fields, respectively), by diffusive addition of individual protein molecules, they continue to grow by colliding into each other and sticking or flocculating. This phase occurs at a slower rate. During the final step, aging in a shear filed, the precipitate particles repeatedly collide and stick, then break apart, until a stable mean particle size is reached, which is dependent upon individual proteins. The mechanical strength of the protein particles correlates with the product of the mean shear rate and the aging time, which is known as the Camp number. Aging helps particles withstand the fluid shear forces encountered in pumps and centrifuge feed zones without reducing in size.

6.5 TYPES OF PRECIPITATION

Proteins are different than most other molecules. The physical structure is just as important as the chemical structure. If the structure of the protein changes, the protein does not have the activity you want.

6.5.1 Isoelectric Precipitation

The isoelectric point (pI) is the pH of a solution at which the net primary charge of a protein becomes zero. At a solution pH that is above the pI the surface of the protein is predominantly negatively charged and therefore like-charged molecules will exhibit repulsive forces. Likewise, at a solution pH that is below the pI, the surface of the protein is predominantly positively charged and repulsion between proteins occurs. However, at the pI the negative and positive charges cancel, repulsive electrostatic forces are reduced and the dispersive forces predominate. The dispersive forces will cause aggregation and precipitation. The pI of most proteins is in the pH range of 4-6. Mineral acids, such as hydrochloric and sulphuric acid are used as precipitants. The greatest disadvantage to isoelectric point precipitation is the irreversible denaturation caused by the mineral acids. For this reason isoelectric point precipitation is most often used to precipitate contaminant proteins, rather than the target protein. The precipitation of casein during cheesemaking, or during production of sodium caseinate, is an isoelectric precipitation.

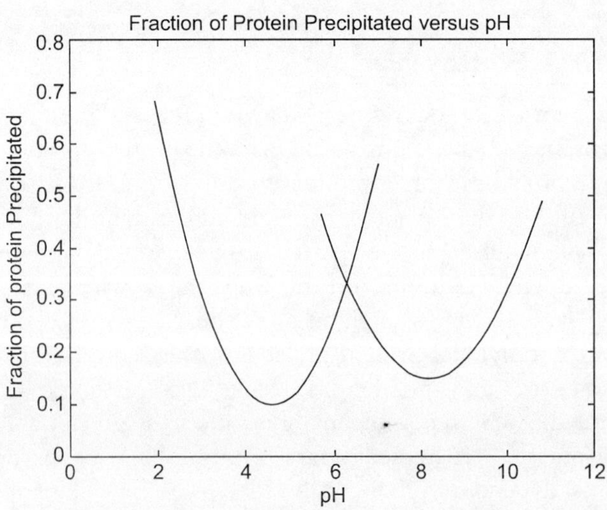

Fig. 2 Fraction of protein precipitated versus pH

Low pH can also precipitate a protein isoelectrically, i.e. as the net charge goes to zero, the protein, and others of the same pI, associate and precipitate, without necessarily denaturing. If the protein of interest can be precipitated fairly quantitatively without denaturing, and then can be redissolved at another pH and is active, one has a good means of purifying it. All precipitation methods, except those depending on ionic interaction with oppositely charged material, are most effective at the isoelectric point of the protein. About 4/5 of proteins are negatively charged at neutral pH and generally have pIs between 4 and 6; the other 1/5 are positively charged at neutral pH, have basic pIs and are not likely to be precipitable at their isoelectric point. Note that the isoelectric point of a protein is unrelated to the pH optimum of its biological activity.

6.5.2 Precipitation by Addition of Salts

Salting out of proteins, particularly by use of ammonium sulphate, is one of the best known and used methods of purifying and concentrating enzymes, particularly at the laboratory scale. Increases in the ionic strength of the solution cause a reduction in the repulsive effect of like charges between identical molecules of a protein. It also reduces the forces holding the solvation shell around the protein molecules. When these forces are sufficiently reduced, the protein will precipitate; hydrophobic proteins precipitating at lower salt concentrations than hydrophilic proteins.

Ammonium sulphate is convenient and effective because of its high solubility, cheapness, lack of toxicity to most enzymes and its stabilizing effect on some enzymes. Its large-scale use, however, is limited as it is corrosive except with stainless steel, it forms dense solutions presenting problems to the collection of the precipitate by centrifugation, and it may release gaseous ammonia, particularly at alkaline pH. The practice of using ammonium sulphate precipitation is more straightforward than the theory. Reproducible results can only be obtained provided the protein concentration, temperature and pH are kept constant. The concentration of the salt needed to precipitate an enzyme will vary with the concentration of the enzyme. However, fractionation of protein mixtures by the stepwise increase in the ionic strength can be a very effective way of partly purifying enzymes.

The solubility of an enzyme can be described by the equation

$$\log S = K_{\mathbf{intercept}} - K_{\mathrm{salt}} T$$

where, S is the enzyme solubility, $K_{\mathbf{intercept}}$ is the intercept constant, K_{salt} is the salting out constant and **T** is the ionic strength which is proportional to the concentration of a precipitating salt. $K_{\mathbf{intercept}}$ is independent of the salt used but depends on the pH, temperature, enzyme and the other components in the solution. K_{salt} depends on both the enzyme required and the salt used but is largely independent of other factors. This equation may also be used to give the minimum salt concentration necessary before enzyme will start to precipitate; the concentration change necessary to precipitate the enzyme varying according to the magnitude of the salting out constant.

Some enzymes do not survive ammonium sulphate precipitation. Other salts may be substituted but the more favoured alternative is to use organic solvents such as methanol, ethanol, propan-2-ol and acetone. These act by reducing the dielectric of the medium and consequently reducing the solubility of proteins by favouring protein-protein rather than protein-solvent interactions. Organic solvents are not widely used on a large scale because of their cost, their flammability, and the tendency of proteins to undergo rapid denaturation by these solvents if the temperature is allowed to rise much above 0°C.

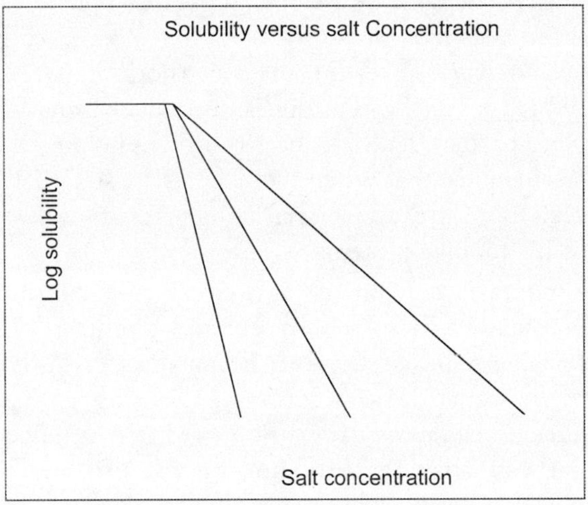

Fig. 2 Solubility versus salt concentration

6.5.3 Precipitation by Addition of Organic Solvents

When large amounts of a water-miscible solvent such as ethanol or acetone are added to a protein solution, proteins precipitate out. The conventional wisdom is that this is due to decrease of the dielectric constant, which would make interactions between charged groups on the surface of proteins stronger. However, it is found that ethanol does not decrease the dielectric constant of water much; indeed 20% EtOH at –5° has the same dielectric constant as water at 20°, also ethanol associates with water much more strongly than do proteins, so that its real effect is to dehydrate protein surfaces, which then associate by van der Waals forces. Removal of water molecules from around charged groups would also deshield them and allow charge interactions to occur more strongly, if you have areas of opposite charge on the surfaces of two proteins. Salts tend to bind to protein surfaces and make them less isoelectric, and therefore tend to mess up ethanol precipitation, which should be carried out at low salt.

Addition of miscible solvents such as ethanol or methanol to a solution may cause proteins in the solution to precipitate. The solvation layer around the protein will decrease as the organic solvent progressively displaces water from the protein surface and binds it in hydration layers around the organic solvent molecules. With smaller hydration layers, the proteins can aggregate by attractive electrostatic and dipole forces. Important parameters to consider are temperature, which should be less than 0°C to avoid denaturation, pH and protein concentration in solution. Miscible organic solvents decrease the dielectric constant of water, which in effect allows two proteins to come close together. At the isoelectric point the relationship between the dielectric constant and protein solubility is given by:

$$\log(S) = k/e^2 + \log S^0,$$

where; S^0 is an extrapolated value of S; **e** is the dielectric constant of the mixture; and **k** is a constant that relates to the dielectric constant of water.

A related method is precipitation with polyethylene glycol, at low concentrations, 5 to 15%. It probably works the same way, by competing with the protein for water, but is less likely to inactivate the protein and does not require such low temperatures.

Mainly ethanol and acetone is used for precipitation. Other solvent such as methanol, n-propanol, iso-propanol, and dioxin is used. Methanol is less effective because of its higher dielectric constant. Long chain alcohols are more denaturing than short ones. Addition of neutral salts to an aqueous-organic solvent solution mixture increases salting-in effect in comparison salting-out effect, thus favors precipitation and it also modifies the value of constant K.

Advantages of using organic solvent:

1. Precipitation depends on the molecular size of the solute; the larger the solute the lower is the concentration of organic solvent required to precipitate it.
2. Organic solvent provides variation in dielectric constant along with pH, temperature and ionic strength which is helpful in protein precipitation.
3. Addition of neutral salts to aqueous-organic solvent mixture produces a salting-in effect and modifies the value of constant K. At decreased dielectric constant salting-in effect increases than salting-out thus favors precipitation

The main disadvantage of organic solvent is their denaturing effect on proteins. In addition they are inflammable necessitating the use of costly flame proof equipment in large scale operations.

6.5.4 Precipitation by Addition of Non-ionic Polymers

Polymers, such as dextrans and polyethylene glycols, are frequently used to precipitate proteins because they have low flammability and are less likely to denature biomaterials than isoelectric precipitation. These polymers in solution attract water molecules away from the solvation layer around the protein. This increases the protein-protein interactions and enhances precipitation. For the specific case of polyethylene glycol, precipitation can be modeled by the equation:

$$\ln(S) + PS = X - aC$$

where; **S** is the enzyme solubility; **C** is the polymer concentration; **P** is a protein-protein interaction coefficient; and **a** is a protein-polymer interaction coefficient and

$$x = (M\mu_i - M\mu_i)RT$$

where; μ is the chemical potential of component **i**, **R** is the universal gas constant; and **T** is the absolute temperature.

When two polymers, typically dextran (an α-1, 6-linked glucose polymer, with 5% α-1, 3 links, produced by the bacterium Leuconostoc) and polyethylene glycol are dissolved in water in appropriate proportions, say 6% PEG and 8% dextran overall, two phases develop, a dextran phase on the bottom and a PEG-rich phase on top. Proteins distribute themselves between the two phases. PEG and high concentrations of salts (phosphate, citrate) can similarly generate two phases. PEG with affinity ligands attached, so that proteins binding to the ligand go into the upper phase. The upper phase is then extracted with a high salt solution, which not only generates a lower phase into which proteins go, but loosens the specific binding to the affinity ligand.

6.5.5 Flocculation by Polyelectrolytes

Alginate, carboxymethycellulose, polyacrylic acid, tannic acid and polyphosphates can form extended networks between protein molecules in solution. The effectiveness of these polyelectrolytes depends on the pH of the solution. Anionic polyelectrolytes are used at pH values less than the isoelectric point. Cationic polyelectrolytes are at pH values above the pI. It is important to note that an excess of polyelectrolytes will cause the precipitate to dissolve back into the solution. An example of polyelectrolyte flocculation is the removal of protein cloud from beer wort using irish moss.

6.6 PRECIPITATION REACTORS

There are numerous industrial scaled reactors than can be used to precipitate large amounts of proteins, such as recombinant DNA polymerases from a solution.

6.6.1 Batch Reactors

Batch reactors are the simplest type of precipitation reactor. The precipitating agent is slowly added to the protein solution under mixing. The aggregating protein particles tend to be compact and regular in shape. Since the particles are exposed to a wide range of shear stresses for a long period of time, they tend to be compact, dense and mechanically stable.

6.6.2 Tubular Reactors

In Tubular Reactors, feed protein solution and the precipitating reagent are contacted in a zone of efficient mixing then fed into long tubes where precipitation takes place. The fluids in volume elements approach plug flow as they move though the tubes of the reactor. Turbulent flow is promoted through wire mesh inserts in the tube. The tubular reactor does not require moving mechanical parts and is inexpensive to build. However, the reactor can become impractically long if the particles aggregate slowly.

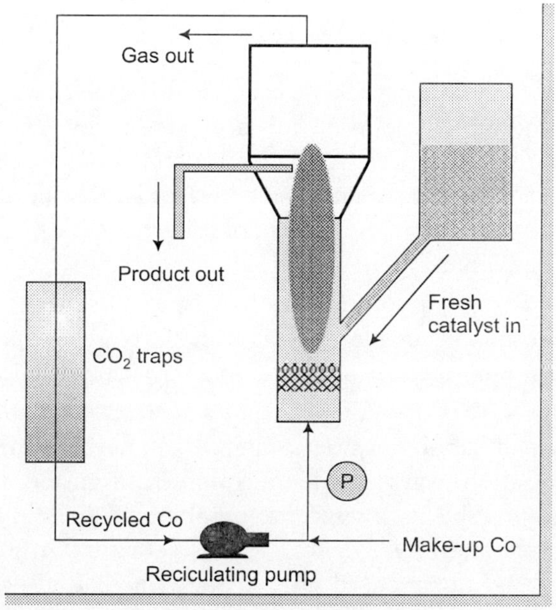

Fig. 3 Tubular Reactor

6.6.3 Continuous Stirred Tank Reactors (CSTR)

This reactor is useful for substrate solutions of high viscosity and for immobilized enzymes with relatively low activity. However, a problem that arises is that an immobilized enzyme tends to decompose upon physical stirring.

The batch system is generally suitable for the production of rather small amounts of chemicals.

A **continuous stirred tank** reactor is shown below:

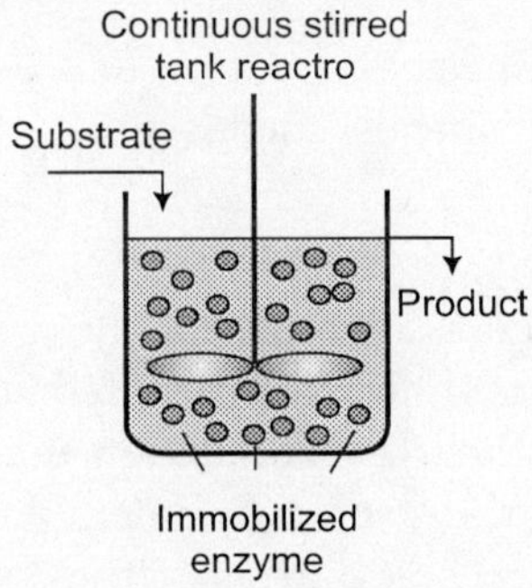

CSTR reactors run at steady state with a continuous flow of reactants and products in a well-mixed tank. Fresh protein feed contacts slurry that already contains precipitate particles and the precipitation reagents.

6.7 SCALE-UP OF PRECIPITATION

Scale-up simply means increasing of scale which is about ten or more times the laboratory scale. Scale-up is important for large production level because it gives the information about process and favorable conditions for precipitation process. The aim of scaling is to maintain kinetics for large scale precipitation same as in small scale precipitation. The steps of relevance in large scale level by appropriate power per volume are:

MIXING – Mixing of a solution at high speed does not produce a homogenous mixture at molecular level. This principle is used for precipitation. When a homogenous mixture does not form, solute precipitates out.

NUCLEATION – The transformation of a metastable material by the growth of initially small fluctuations (e.g., composition fluctuations in the case of spinodal decomposition) was treated as a continuous transformation. Some metastable systems are stable with respect to infinitesimal fluctuations, but are unstable to a perturbation in the form of a finite fluctuation. In such systems, the transformation does not proceed until a finite fluctuation occurs. If the necessary fluctuation is so large that its probability of it occurring in any observable time is effectively zero, then the system is kinetically 'frozen' in its metastable state. Many, if not most, engineering materials are in such non-equilibrium states.

Each of these finite fluctuations is called a *nucleus*; the metastable medium is called the *matrix*. The minimum size nucleus required for continued

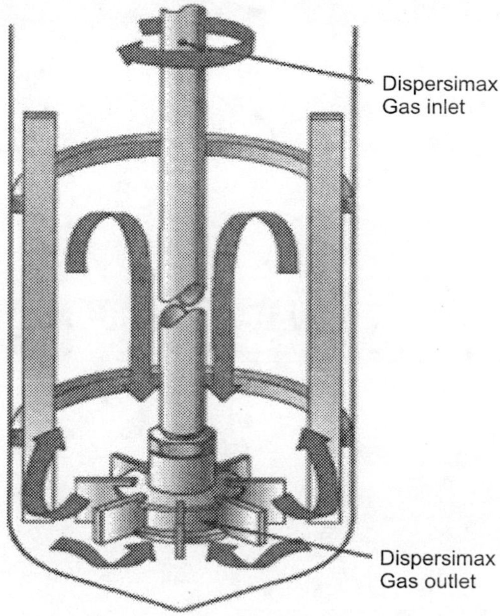

Fig. 5 Continuous stirred tank reactor.

spontaneous growth is called a *critical nucleus* and all the fluctuations smaller than the critical nucleus are called *sub-critical nuclei.* The process by which the sub-critical nuclei change their size by the (energetically favorable) loss of its atoms to the matrix and the improbable attachment of new matrix atoms onto the sub-critical nucleus is called *nucleation.* This nucleation is the part of the process where small particles appear and begin to grow. This process is mainly carried out for biological system

7

CHAPTER

CHROMATOGRAPHY

7.1 WHAT IS CHROMATOGRAPHY?

Well, quite simply, it is a broad range of physical methods used to separate and or to analyze complex mixtures. The components to be separated are distributed between two phases: a *stationary phase* bed and a *mobile phase* which percolates through the stationary bed.

7.2 HOW DOES IT WORK?

A mixture of various components enters a chromatography process, and the different components are flushed through the system at different rates. These differential rates of migration as the mixture moves over adsorptive materials provide separation. Repeated sorption/desorption acts that take place during the movement of the sample over the stationary bed determine the rates. The smaller the affinity a molecule has for the stationary phase, the shorter the time spent in a column.

7.3 WHY CHROMATOGRAPHY IS SO SPECIAL?

In any chemical or bioprocess industry, the need to separate and purify a product from a complex mixture is a necessary and important step in the production line. Today, there exists a wide market of methods in which industries can accomplish these goals. Chromatography is a *very* special separation process for a multitude of reasons! First of all, it can separate complex mixtures with great precision. Even very similar components, such as proteins that may only vary by a single amino acid, can be separated with chromatography. In fact, chromatography can purify basically any soluble or volatile substance if the right adsorbent material, carrier fluid, and operating conditions are employed. Second, chromatography can be used to separate delicate products since the conditions under which it is performed are not typically severe. For these reasons, chromatography is quite well suited to a variety of uses in the field of biotechnology, such as separating mixtures of proteins.

7.4 CHROMATOGRAPHY – EQUIPMENT

The Column

Although there are other types of chromatography (e.g. paper and thin layer), most modern applications of chromatography employ a *column*. The column is where the actual separation takes place. It is usually a glass or metal tube of sufficient strength to withstand the pressures that may be applied across it. The column contains the *stationary phase*. The *mobile phase* runs through the column and is adsorbed onto the stationary phase. The column can either be a *packed bed* or *open tubular* column.

- *Packed Bed Column:* A packed bed column is comprised of a stationary phase which is in granular form and packed into the column as a homogeneous bed. The stationary phase completely fills the column.
- *Open Tubular Column:* An open tubular column's stationary phase is a thin film or layer on the column wall. There is a passage-way through the center of the column.

The figure below will help you to follow along in our discussion.

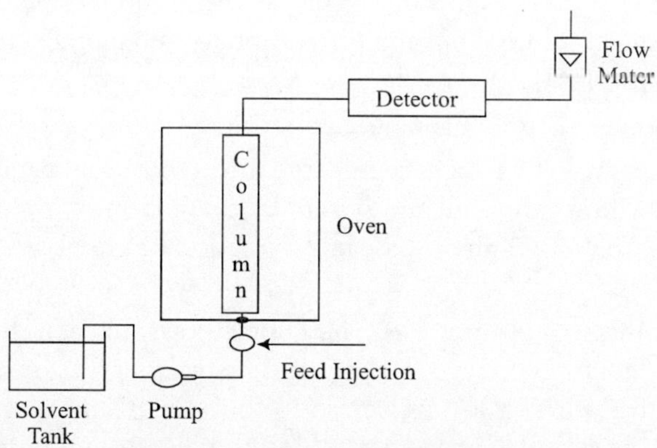

Fig. 1 Chromatography Column

The Mobile and Stationary Phases

The mobile phase is comprised of a solvent into which the sample is injected. The solvent and sample flow through the column together; thus the mobile phase is often referred to as the "carrier fluid." The stationary phase is the material in the column for which the components to be separated have varying affinities. The materials which comprise the mobile and stationary phases vary depending on the general type of chromatographic process being performed.

Gas Chromatography

The mobile phase in gas chromatography is generally an inert gas. The stationaryphase is generally an adsorbent or liquid distributed over the surface of a porous, inert support.

Liquid Chromatography

The mobile phase in liquid chromatography is a liquid of low viscosity which flows through the stationary phase bed. This bed may be comprised of an immiscible liquid coated onto a porous support, a thin film of liquid phase bondedto the surface of a sorbent, or a sorbent of controlled pore size.

7.5 CHROMATOGRAPHY – BASIC OPERATION

Review paragraph. The process of a chromatographic separation takes place within a chromatography column. This column made of glass or metal, is either a packed bed or open tubular column. A packed bed column contains particles which make up the stationary phase. Open tubular columns are lined with a thin film stationary phase. The center of the column is hollow. The mobile phase is typically a solvent moving through the column which carries the mixture to be separated. This can either be a liquid or a gas, depending on the type of process. The stationary phase is usually a viscous liquid coated on the surface of solid particles which are packed into the column as discussed above, although the solid particles can also be taken as the stationary phase. In any case, the partitioning of solutes between the stationary and mobile phases leads to the desired separations.

Let's see a different sketch of the process:

1. **Feed Injection:** The feed is injected into the mobile phase. The mobile phase flows through the system by the action of a pump (older analytical chromatography used capillary action or gravity to move the mobile phase).
2. **Separation in the Column:** As the sample flows through the column, its different components will adsorb to the stationary phase to varying degrees. Those with strong attraction to the support move more slowly than those with weak attraction. This is how the components are separated.
3. **Elution from the Column:** After the sample is flushed or displaced from the stationary phase, the different components will elute from the column at different times. The components with the least affinity for the

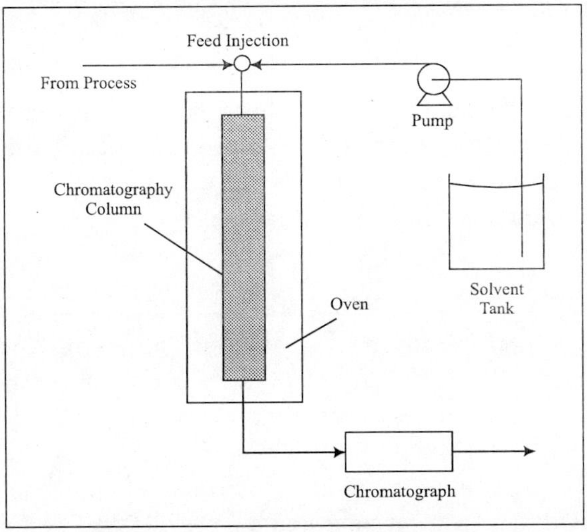

Fig. 2 Chromatography Operation

stationary phase (the most weakly adsorbed) will elute first, while those with the greatest affinity for the stationary phase (the most strongly adsorbed) will elute last.

4. **Detection:** The different components are collected as they emerge from the column. A detector analyzes the emerging stream by measuring a property which is related to concentration and characteristic of chemical composition. For example, the refractive index or ultra-violet absorbance is measured.

Example: The figure below shows a simple separation by chromatography. A continuous flow of solvent carries a solution of solutes A and B down a column.

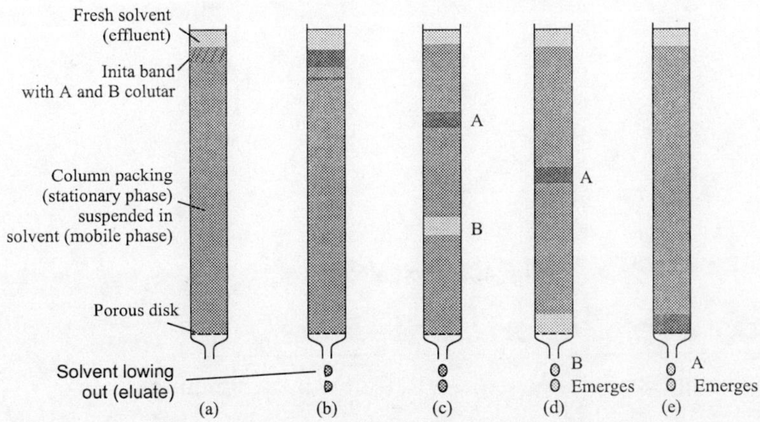

Fig. 3 Example

(a) As the solvent carries the two solutes down the column, we begin to see some separation of the solution. (b) At some later point in time, it can be seen that solute B is moving at a much faster rate than A. (c) In (d), solute B emerges first, while solute A finally emerges in (e). Thus, solute A has a greater affinity for the stationary phase than solute B. By varying the pH of the solvent or temperature of the column, the output of the column can be significantly altered, such as the timing of when individual species emerge.

7.6 CHROMATOGRAPHY—THE CHROMATOGRAM

Since the sample is separated in the column, different peaks on the chromatogram correspond to different components in the sample mixture. The chromatograms above show the results of separations of protein mixtures by ion exchange chromatography. The lettered peaks correspond to different proteins (A = ovalbumin, B = conalbumin, C = cytochrome c, D = lysozyme). The separation corresponding to the chromatogram on the left was performed at pH 5.85, while the one on the right was performed at pH 6.5. It is evident that operation conditions such as pH and temperature have a significant effect on the output.

Let's consider the output from the detector: the chromatogram. The figure below will help you to follow along in our discussion.

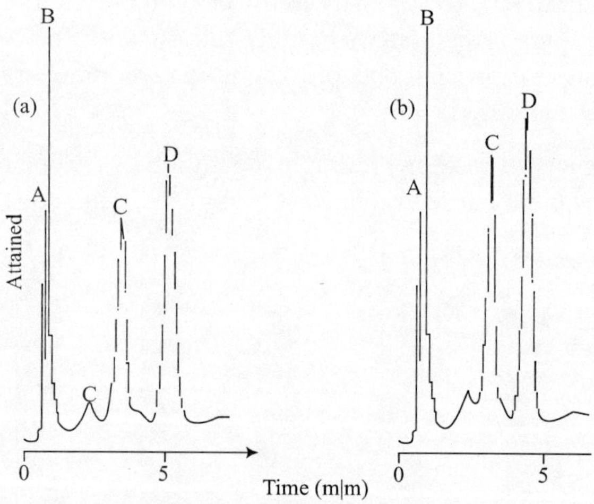

Fig. 4 Chromatogram

7.7 WHAT INFORMATION CAN BE ATTAINED?

The level of complexity of the sample is indicated by the number of peaks which appear.

- Qualitative information about the sample composition is obtained by comparing peak positions with those of standards.
- Quantitative assessment of the relative concentrations of components is obtained from peak area comparisons.
- Column performance is indicated by comparison with standards.

7.8 TYPES OF CHROMATOGRAPHY

7.8.1 Gas Chromatography

Gas chromatography makes use of a pressurized gas cylinder and a carrier gas, such as helium, to carry the solute through the column. The most common detectors used in this type of chromatography are thermal conductivity and flame ionization detectors. There are three types of gas chromatography that will be discussed here: gas adsorption, gas-liquid and capillary gas chromatography.

Gas adsorption chromatography involves a packed bed comprised of an adsorbent used as the stationary phase. Common adsorbents are zeolite, silica gel and activated alumina. This method is commonly used to separate mixtures of gases.

Gas-liquid chromatography is a more common type of analytical gas chromatography. In this type of column, an inert porous solid is coated with a viscous liquid which acts as the stationary phase. Diatomaceous earth is the most common solid used. Solutes in the feed stream dissolve into the liquid phase and eventually vaporize. The separation is thus based on relative volatilities.

Capillary gas chromatography is the most common analytical method. Glass or fused silica comprises the capillary walls which are coated with an absorbent or other solvent. Because of the small amount of stationary phase, the column can contain only a limited capacity. However, this method also yields rapid separation of mixtures.

7.8.2 Liquid Chromatography

There are a variety of types of liquid chromatography. There is liquid adsorption chromatography in which an adsorbent is used. This method is used in large-scale applications since adsorbents are relatively inexpensive. There is also liquid-liquid chromatography which is analogous to gas-liquid chromatography. The three types that will be considered here fall under the category of modern liquid chromatography. They are reverse phase, high performance and size exclusion liquid chromatography, along with supercritical fluid chromatography.

Reverse phase chromatography is a powerful analytical tool and involves a hydrophobic, low polarity stationary phase which is chemically bonded to an inert solid such as silica. The separation is essentially an extraction operation and is useful for separating non-volatile components.

High performance liquid chromatography (HPLC) is similar to reverse phase, only in this method, the process is conducted at a high velocity and pressure drop. The column is shorter and has a small diameter, but it is equivalent to possessing a large number of equilibrium stages.

Size exclusion chromatography, also known as gel permeation or filtration chromatography does not involve any adsorption and is extremely fast. The packing is a porous gel, and is capable of separating large molecules from smaller ones. The larger molecules elute first since they cannot penetrate the pores. This method is common in protein separation and purification.

Supercritical fluid chromatography is a relatively new analytical tool. In this method, the carrier is a supercritical fluid, such as carbon dioxide mixed with a modifier. Compared to liquids, supercritical fluids have solubilities and densities have as large, and they have diffusivities and viscosities quite a bit larger. This type of chromatography has not yet been implemented on a large scale.

7.8.3 Ion Exchange Chromatography

Ion exchange chromatography is commonly used in the purification of biological materials. There are two types of exchange: cation exchange in which the stationary phase carries a negative charge, and anion exchange in which the stationary phase carries a positive charge. Charged molecules in the liquid phase pass through the column until a binding site in the stationary phase appears. The molecule will not elute from the column until a solution of varying pH or ionic strength is passed through it. Separation by this method is highly selective. Since the resins are fairly inexpensive and high capacities can be used, this method of separation is applied early in the overall process.

7.8.4 Affinity Chromatography

Affinity chromatography involves the use of packing which has been chemically modified by attaching a compound with a specific affinity for the desired molecules, primarily biological compounds. The packing material used, called the affinity matrix, must be inert and easily modified. Agarose is the most common substance used, in spite of its cost. The ligands, or "affinity tails", that are inserted into the matrix can be genetically engineered to possess a specific affinity. In a process similar to ion exchange chromatography, the desired molecules adsorb to the ligands on the matrix until a solution of high salt concentration is passed through the column. These causes desorption of the molecules from the ligands, and they elute from the column. Fouling of the matrix can occur when a large number of impurities are present; therefore, this type of chromatography is usually implemented late in the process.

7.9 DIAGRAMATICAL GUIDE TO VARIOUS CHROMAROGRAPHIC TECHNIQUES

Affinity Chromatography

This is the most selective type of chromatography employed. It utilizes the specific interaction between one kind of solute molecule and a second molecule that is immobilized on a stationary phase. For example, the immobilized molecule may be an antibody to some specific protein. When solute containing a mixture of proteins is passed by this molecule, only the specific protein is reacted to this antibody, binding it to the stationary phase. This protein is later extracted by changing the ionic strength or pH.

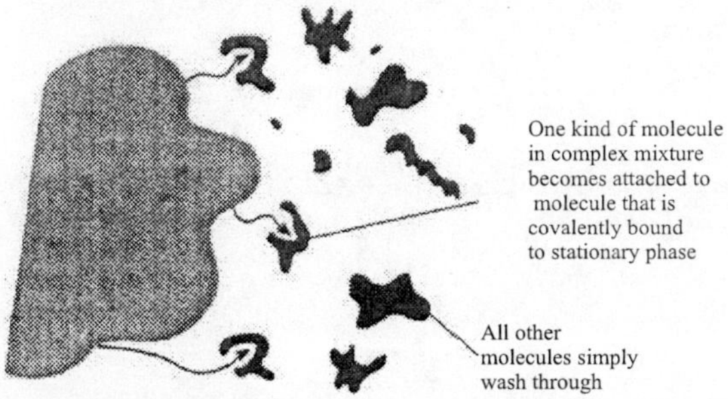

Affinity chromatography

Adsorption Chromatography

Adsorption chromatography is probably one of the oldest types of chromatography around. It utilizes a mobile liquid or gaseous phase that is adsorbed onto the surface of a stationary solid phase. The equilibration between the mobile and stationary phase accounts for the separation of different solutes.

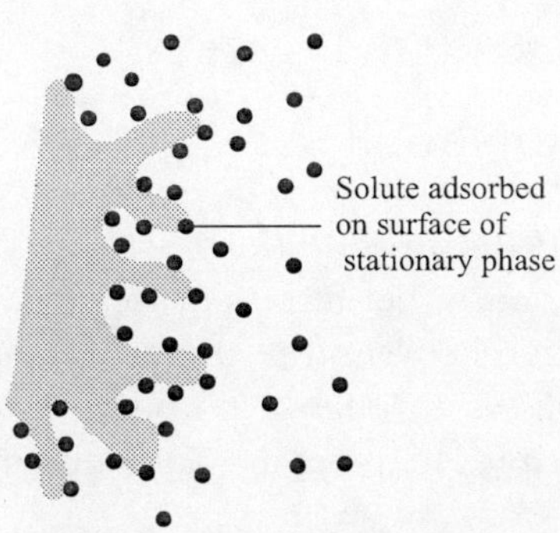

Adsorption chromatography

Ion Exchange Chromatography

In this type of chromatography, the use of a resin (the stationary solid phase) is used to covalently attach anions or cations onto it. Solute ions of the opposite charge in the mobile liquid phase are attracted to the resin by electrostatic forces.

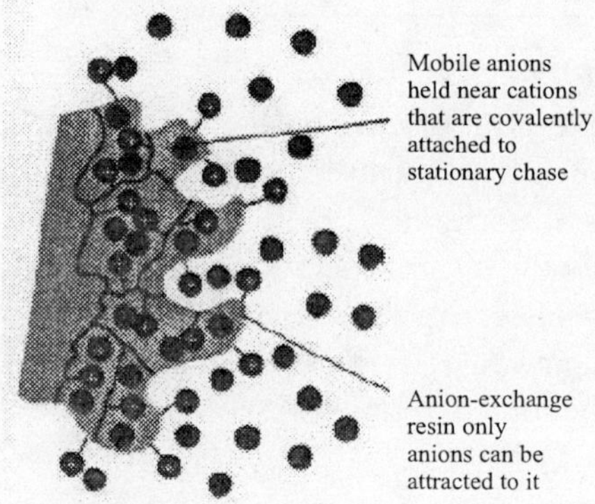

Ion-exchange chromatography

Partition Chromatography

This form of chromatography is based on a thin film formed on the surface of a solid support by a liquid stationary phase. Solute equilibrates between the mobile phase and the stationary liquid.

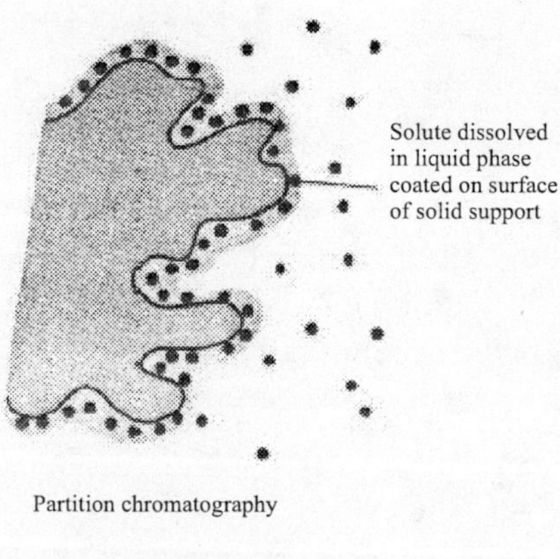

Partition chromatography

Molecular Exclusion Chromatography

Also known as gel permeation or gel filtration, this type of chromatography lacks an attractive interaction between the stationary phase and solute. The liquid or gaseous phase passes through a porous gel which separates the molecules according to its size. The pores are normally small and exclude the larger solute molecules, but allow smaller molecules to enter the gel, causing them to flow through a larger volume. This causes the larger molecules to pass through the column at a faster rate than the smaller ones.

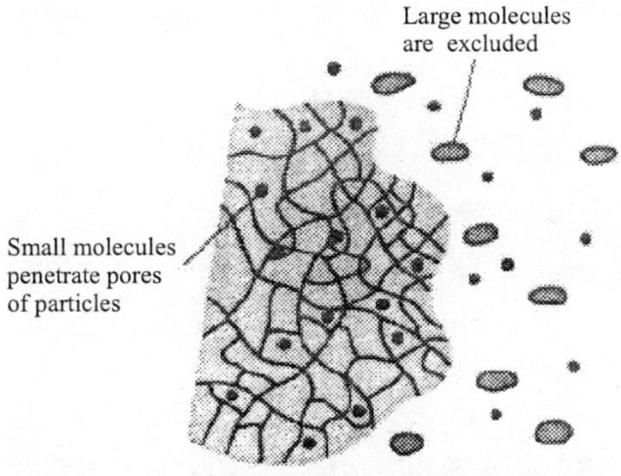

Molecular exclusion chromatography

APPENDIX
High-Performance Liquid Chromatography (HPLC)

INTRODUCTION

High performance liquid chromatography (HPLC) is a form of liquid chromatography to separate compounds that are dissolved in solution. HPLC instruments consist of a reservoir of mobile phase, a pump, an injector, a separation column, and a detector. Compounds are separated by injecting a plug of the sample mixture onto the column. The different components in the mixture pass through the column at different rates due to differences in their partitioning behavior between the mobile liquid phase and the stationary phase.

Principle

High performance liquid chromatography (HPLC) is a separation technique utilizing differences in distribution of compounds to two phases, called stationary phase and mobile phase. The stationary phase designates a thin layer created on the surface of fine particles and the mobile phase designates the liquid flowing over the particles. Under a certain dynamic condition, each component in a sample has different distribution equilibrium depending on solubility in the phases and/or molecular size. As a result, the components move at different speeds over the stationary phase and are thereby separated from each other. This is the principle behind HPLC. The column is a stainless steel (or resin) tube which is packed with spherical solid particles. Mobile phase is constantly fed into the column inlet at a constant rate by a liquid pump. A sample is injected from a sample injector, located near the column inlet. The injected sample enters the column with the mobile phase and the components in the sample migrate through it, passing between the stationary and mobile phases. Compounds move in the column only when is in the mobile phase. Compounds that tend to be distributed in the mobile phase therefore migrate faster through the column while compounds that tend to be distributed in the stationary phase migrate slower. In this way, each component is separated on the column and sequentially elutes from the outlet. Each compound eluting from the column is detected by a detector connected to the outlet of the column.

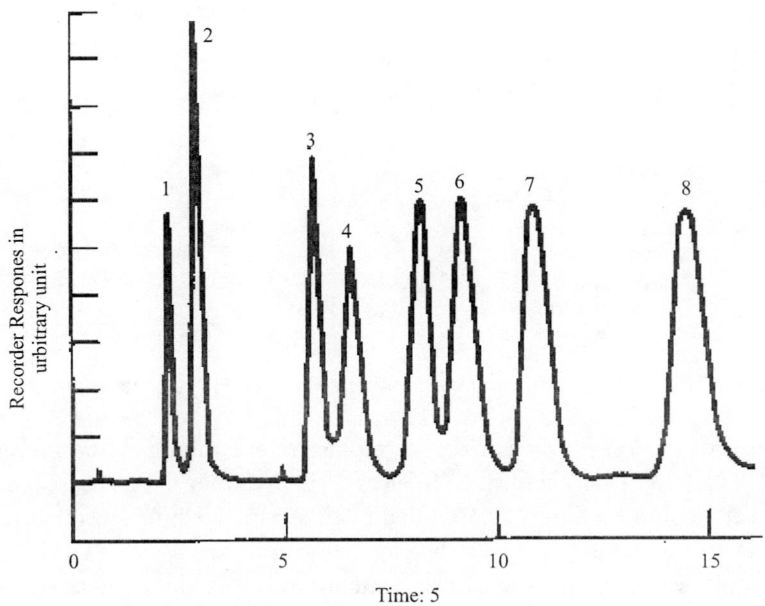

Fig. 5 Chromatogram showing eight distinct peaks, each representing distinct compounds.

When the separation process is monitored by the recorder starting at the time the sample is injected, a graph is obtained. This graph is called a chromatogram (Figure 5). The time required for a compound to elute (called retention time) and the relationship between compound concentration (amount) and peak area depend on the characteristics of the compound. Retention time is therefore used as an index for qualitative determination and peak surface area (or height) as an index for quantitative determination. The retention time of the target compounds and the concentration for each unit of peak area are based on data obtained in advance by analyzing a sample with known quantities of the reference standards. Normally, the reference standards are highly purified target compounds.

A HPLC system is basically composed of 1) a pump, 2) an injector, 3) a column, 4) a column oven and 5) detector.

Instrumentation

Solvents must be degassed to eliminate formation of bubbles. The pumps provide a steady high pressure with no pulsating, and can be programmed to vary the composition of the solvent during the course of the separation. The liquid sample is introduced into a sample loop of an injector with a syringe. When the loop is filled, the injector can be inject the sample into the stream by

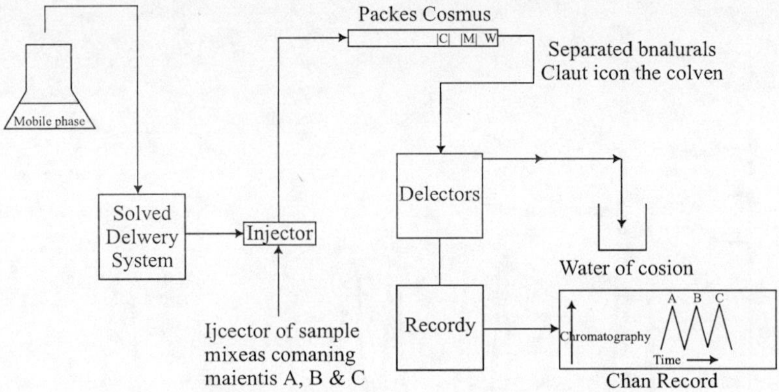

Fig. 6 Block diagram showing the components of an HPLC instrument

placing the sample loop in line with the mobile phase tubing. The different types of HPLC columns are described in a separate document. The presence of analytes in the column effluent is recorded by detecting a change in refractive index, UV-VIS absorption at a set wavelength, fluorescence after excitation with a suitable wavelength, or electrochemical response. Mass spectrometers can also be interfaced with liquid chromatography to provide structural information and help identify the separated analytes.

Fig. 7 Photograph of an HPLC apparatus.

Liquid Chromatography Columns and Stationary Phases

Columns

Conventional liquid chromatography uses plastic or glass columns that can range from a few centimeters to several meters. The most common lengths are 10-100 cm, with the longer columns finding use for preparative-scale separations. High-performance liquid chromatography (HPLC) columns are stainless steel tubes, typically of 10-30 cm in length and 3-5 mm inner diameter.

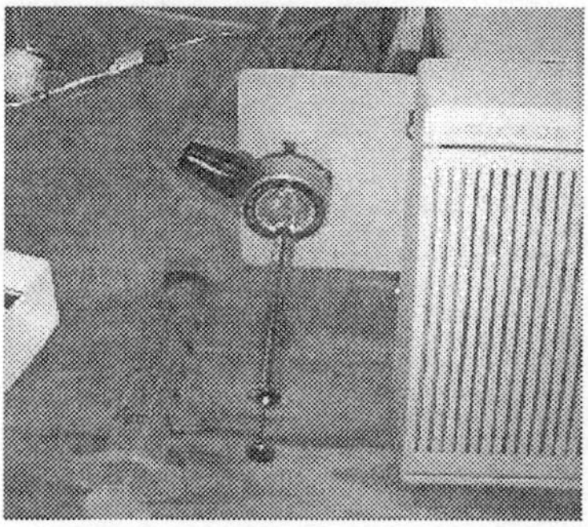

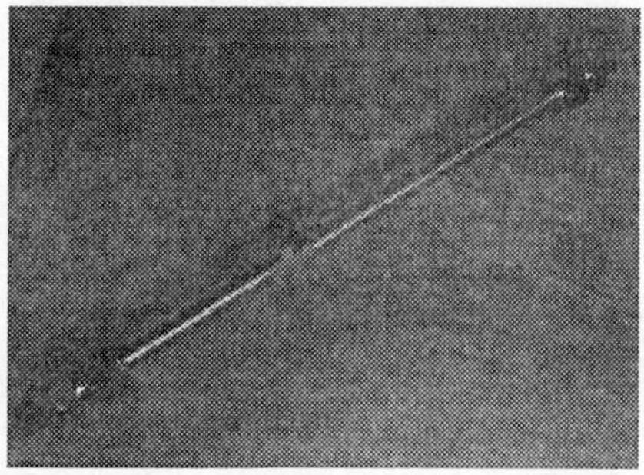

Fig. 8 Close up of an injector (left) and Picture of an HPLC column (right)

Short, fast analytical columns, and guard columns, which are placed before an analytical column to trap junk and extend the lifetime of the analytical column, are 3-10 cm long.

Partition Chromatography

Stationary Phases

In partition chromatography the stationary phase is bonded to inert particles of 3-10 ?m of diameter, with the smaller sizes, 3-5 ?m, being used in analytical columns, and the larger particles being used in preparative-scale HPLC. Analytes separate as they travel through the column due to the differences in their partitioning between the mobile phase and the stationary phase.

Reverse-phase partition chromatography uses a relatively nonpolar stationary phase and a polar mobile phase, such as methanol, acetonitrile, water, or mixtures of these solvents. The most common bonded phases are n-octyldecyl (C_{18}) and n-decyl (C_8) chains, and phenyl groups. Reverse-phase chromatography is the most common form of liquid chromatography, primarily due to the wide range on analytes that can dissolve in the mobile phase.

In reverse phase chromatography, the packing is nonpolar and the solvent is polar with respect to the sample. Retention is the result of the interaction of the nonpolar components of the solutes and the nonpolar stationary phase. Typical stationary phases are nonpolar hydrocarbons, waxy liquids or bonded hydrocarbons (such as C_{18}, C_8, C_4, etc.) and the solvents are polar aqueous-organic mixtures such as methanol-water or acetonitrile-water.

The C_{18}, C_8, and phenyl bonded phases are most often used in the reverse phase mode. It has been estimated that 60-90% of all analytical LC separations are done on bonded phases in the reverse phase mode. Bonded phases made by covalently bonding a molecule onto a solid stationary phase are intended to prepare "liquid coatings" which will be permanent. Silica is a reactive substrate to which various functionalities can be attached or bonded. The functionalities most widely bonded to silica are the alkyl (C_{18} and C_8), aromatic phenyl, and cyano and amino groups.

General characteristics of reversed phase chromatography

- Broad scope which allows sample types with a wide range of polarities and molecular weights to be separated.
- General rapidity of mobile phase column equilibration during methods development and gradient regeneration.
- General ease of use.
- Applicability to separation of ionic or ionizable compounds by manipulating secondary chemical equilibrium such as ionization control and ion pairing in the aqueous mobile phase.
    - o Buffering the mobile phase in the pH range from 2 to 5 with one of the common buffers, the ionization of the weak acids can be suppressed or controlled allowing them to be retained in their neutral form. Similarly weak bases can be retained in their neutral form at pH 7-7.5.
    - o For strong acids and bases ionization control cannot be employed because the stability of alkyl bonded phases is diminished below pH 2 and above pH 7.5. Highly hydrophilic weak acids and bases often remain difficult to retain with ionization control. In such cases ion pair reversed phase chromatography can be used. In this method, counterions (species of opposite charge to the solutes) thereby regulate the retention. Typically alkyl amines or tetra alkyl amines are added to ion pair with acids whereas alkyl sulfates, sulfonates, or phosphates are used to ion pair with bases. The technique is an alternative to ion exchange chromatography for analysis of ionic compounds.
- The possibility of special selectivity such as structural or steric are achievable by specific mobile phase additives:
    - o Metal ions are capable of binding to organic compounds in a very selective method which is used for ligand exchange chromatography. The selectivity generated in these metal ion phase systems is based in part on differences of the solute (ligand) binding strength to the metal ion. An alternate approach is the addition of various chelating agents (4-dodecyldiethylene-triamine - C_{12} dien) in combination with a metal ion. The type and strength of the metal chelate complex-solute binding can be greatly varied depending upon the chemical environment surrounding the metal ion as determined by the chelating agent added.

Normal-phase partition chromatography uses a polar stationary phase and a nonpolar organic solvent, such as n-hexane, methylene chloride, or chloroform, as the mobile phase. The stationary phase is a bonded siloxane with a polar functional group. The most common functional groups in order of increasing polarity are:

cyano:	$-C_2H_4CN$
diol:	$-C_3H_6OCH_2CHOHCH_2OH$
amino:	$-C_3H_6NH_2$
Dimethylamino:	$-C_3H_6N(CH_3)_2$

Adsorption Chromatography

The stationary phase in adsorption chromatography are silica or alumina particles. Analytes are separated due to their varying degree of adsorption onto the solid surfaces. The main advantage of adsorption chromatography is in separating isomers, which can have very different physisorption characteristics due to steric effects in the molecules.

Chiral Stationary Phases

Separating enantiomeric mixtures is very challenging because the optical isomers are identical chemically. Chiral analytes can be separated by using a chiral stationary phase. One optical isomer of a chiral molecule is usually bonded to a polymer, which is then coated onto silica packing material. Separations occur because the chiral analytes will interact with the one isomer on the stationary phase differently.

8

CHAPTER

ELECTROPHORESIS

8.1 INTRODUCTION

When subjected to an electric field in an electrophoresis chamber, a charged molecule will move toward the oppositely charged electrode. The rate at which a charged molecule moves in a stable field (its electrophoretic mobility) depends upon two factors specific to the molecule: one is the sign and magnitude of its net electrical charge, and the other is its size and shape. All other factors being equal, if molecules are of equal size the one with higher net charge will move faster in an applied electrical field due to the molecular sieving properties of the solid medium. It also follows that small molecules will move faster than large ones of the same net charge.

Succinctly, if an electric field is applied to a solution containing a protein molecule, the protein will migrate at a rate that depends on its net charge and on its size and shape. This technique, known as electrophoresis, was originally used to separate mixtures of proteins either in free aqueous solution or in solutions held in a solid porous matrix such as starch.

Proteins usually have a net positive or negative charge that reflects the mixture of charged amino acids they contain. Although there are exceptions in which the shape of a molecule may increase or decrease its frictional drag and cause atypical migration behavior, these general principles underlie all electrophoretic separations. Most electrophoretic separations are not conducted in free solution but rather in a stable supporting medium, such as a gel. The most popular in reseach laboratories is a polymerized and crosslinked form of acrylamide. Separation on polyacrylamide gels, commonly referred to as polyacrylamide gel electrophoresis (PAGE), may be used for analysis of proteins or nucleic acids.

8.2 PRINCIPLE & THEORY

Electrophoresis, like centrifugation, is a hydrodynamic technique. A charged particle (i.e., molecule) in an electric field experiences a force that is proportional to the potential difference (E), or voltage, of the electric field and inversely proportional to the distance (d) between the electrodes. (The potential difference divided by the distance (E/d) is referred to as the field strength.) The force will also be proportional to the net charge of the molecule (q). Therefore, the force experienced by the molecule can be expressed by the following equation:

$$\mathbf{F = Eq/d}$$

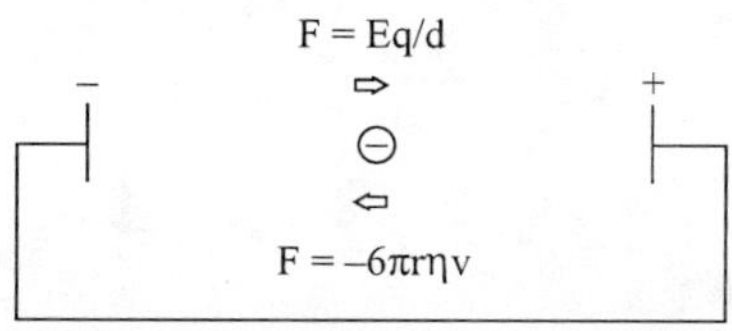

This force will by opposed by a frictional force (= fv) where f is a frictional coefficient and v is the velocity of the particle. The frictional coefficient depends on the size (e.g., r = radius) and shape of the molecule and the viscosity (η) of the medium. For example, in the case of a sphere the frictional force is:

$$\mathbf{F_f = 6\pi r \eta v}$$

A particle will move at a velocity (v) so that these two forces are equal, therefore:

$$\mathbf{6\pi r \eta v = Eq/d} \text{ or solving for v}$$

$$\mathbf{v = Eq/d\ 6\pi r \eta}$$

This equation indicates that the mobility (i.e., velocity) of a molecule in an electric field is proportional to the electric field (E/d), or more simply the applied voltage, and the net charge of the molecule. The mobility is inversely proportional to a frictional coefficient (i.e., size and shape of the molecule and the viscosity of the medium), as indicated by the following equation:

Mobility = (applied voltage) (net charge) ÷ (friction coefficient)

Therefore, it is possible to derive information about the charge, size and shape of a molecule by its mobility in an electric field.

8.3 ELECTROPHORESIS EQUIPMENT

Equipment to conduct gel electrophoresis is relatively simple. They consist of a mold to form the gels, an apparatus to hold the gel and contain buffers, and a power supply capable of delivering the required voltage or current. There are many types of apparati for carrying out electrophoresis depending on the application. Gels can be either in a vertical or horizontal configuration. Polyacrylamide gels are run in a vertical fashion and agarose gels tend to be run in a horizontal position.

Gels can either be formed as cylinders by using glass tubing as a mold (often called tube gels) or formed as rectangular slabs. These slab gels are formed by polymerizing the acryamide solution between glass plates separated by spacers (Figure 1). Typically the gel is 0.75-1.5 mm thick. At the top a 'comb' is used to form sample wells. Slab gels allow multiple samples to be compared on the same gel, thus eliminating gel-to-gel variations.

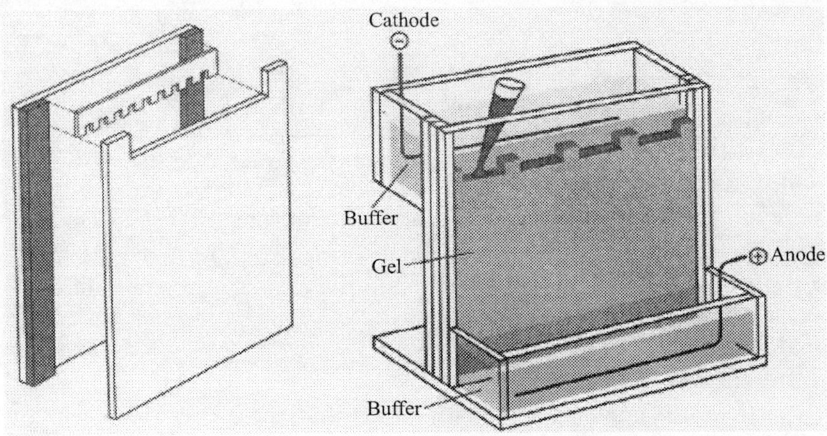

Fig. 1 Common electrophoretic apparatus

The formed gel is placed into the apparatus to that the top and bottom of the gel are in contact with chambers containing buffer (Figure 1). These chambers contain electrodes which are connected to a power supply. Thus an electric field is generated across the gel when a voltage is applied. The buffer in the chambers is generally different that the buffer making up the gel for protein electrophoresis and in some applications the buffers in the lower and upper chambers may be different. In most applications the buffers are such that the protein has a negative charge and therefore the anode (positive pole) will be in the lower chamber and the cathode (negative pole) will be in the upper chamber. However, there are applications in which the proteins of interest may be positively charged and therefore the electrodes will be reversed.

Discontinuous or "disc" electrophoresis: The Laemmli discontinuous buffers are extensively used in gel electrophoresis. Discontinuous gels consist of two distinct gel regions referred to as stacking gel and separating gel (see Table) and a Tris-glycine tank buffer. The stacking gel has a lower acrylamide concentration, a lower pH and a lower ionic strength than the separating gel.

The lower ionic strength of the stacking gel results in a greater local electric field strength than in the separating gel. The field strength difference combined with the lower acrylamide concentration results in proteins having a higher mobility in the stacking gel than in the separating gel. In addition, the glycine in the tank buffer has a higher mobility in the separating gel than in the stacking gel because of the pH differences. Therefore, proteins will migrate faster than the glycine in the stacking gel. When proteins reach the separating gel their mobility is decreased because of the increased acrylamide concentration and decreased field strength, whereas the increase in pH results in glycine having a higher mobility. All of these factors result in the proteins becoming compressed at the interface between the two gels and thus increasing resolution (Figure 2).

Composition of Laemmli Gels		
	Stacking Gel	Separating Gel
Acrylamide	3-4.5%	6-20%
pH	6.8	8.8
Ionic Strength	0.125 M Tris	0.375 M Tris

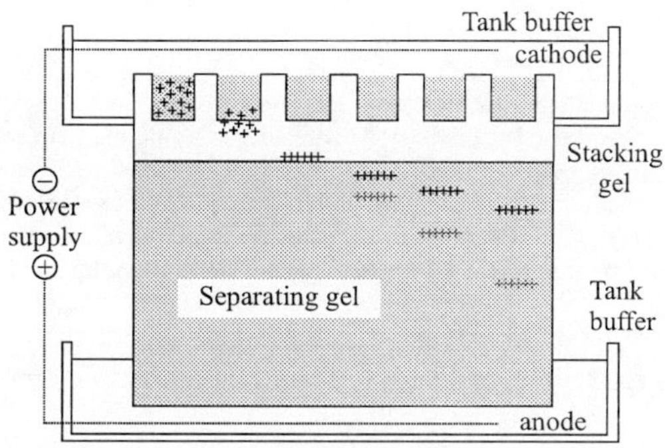

Fig. 2 A vertical gel apparatus used for separating proteins

Resolution in non-discontinuous electrophoresis depends partially on the volume of the sample. However, stacking also occurs at the interface of the sample and gel, especially if a high voltage is applied.

8.4 GEL ELECTROPHORESIS

Electrophoresis of macromolecules can be carried out in solution. However, the ability to separate molecules is compromised by their diffusion. Greater resolution is achieved if electrophoresis is carried out on semi-solid supports such as polyacrylamide or agarose gels. Gels are formed by cross-linking polymers in aqueous medium. This will form a 3-dimensional meshwork which the molecules must pass through. Polyacrylamide is a common gel for protein electrophoresis whereas agarose is more commonly used for nucleic acids. Agarose gels have a larger pore size than acrylamide gels and are better suited for larger macromolecules. However, either type of gel can be applied to either nucleic acids or proteins depending on the application.

Gels are formed from long polymers in a cross-linked lattice (Figure 3). The spaces between the polymers are the pores. Higher concentrations of the

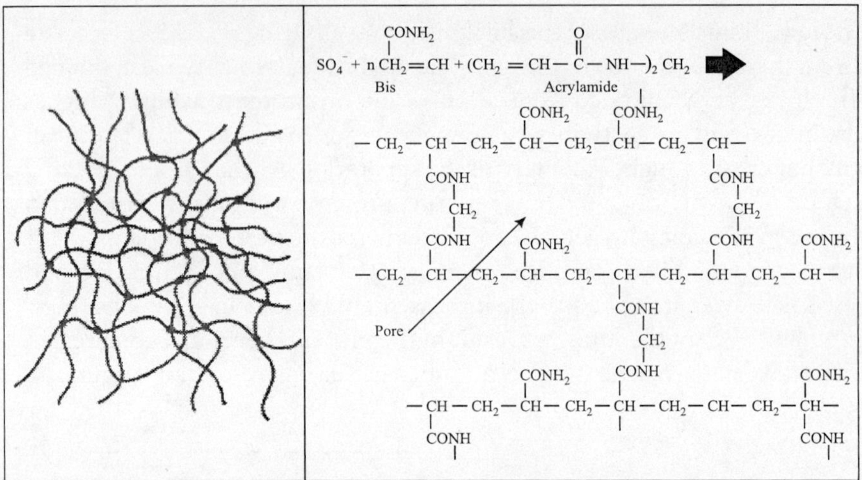

Diagramatic representation of gels. Gels are form by cross-linking polymers into a 3-dimensional mesh-work or lattice (left). This results in a pourous semisolid material that solutions can pass through. In other words the open spaces are filled with solvent. Increasing the concentration of the polymer will resulting in smaller 'pores'. Agarose gels are formed by heating the desired concentration of agarose in an aqueous buffer. After the agarose dissolves, the solution is cooled and the gel forms. Polyacrylamide gels are formed by chemically crossing acrylamide and bis-acrylamide using a free-radical such as persulfate (right).

Fig. 3 Cross linked lattice of long polymers formed in polyacrylamide gels

polymer will result in smaller average pore sizes. Polyacrylamide gels are formed by covalently cross-linking acrylamide monomers with bis-acrylamide with a free radical like persulfate (SO_4). The cross-linking of the acrylamide polymers results in 'pores' of a defined size. The total acrylamide concentration and the ratio of bis-acrylamide to acrylamide will determine the average pore size. The polyacrylamide solution is poured into a mold and polymerized. This mold can be a cylindrical tube, but is usually a 'slab' poured between two glass plates (see below).

Since the gel is solid with respect to the mold, all molecules are forced through the gel. Smaller molecules will be able to pass through this lattice more easily resulting in larger molecules having a lower mobility than smaller molecules. In other words, the gel acts like a molecular sieve and retains the larger molecules while letting the smaller ones pass through. (This is opposite of gel filtration where the larger molecules have a higher mobility because they to not enter the gel.) Therefore, the frictional coefficient is related to how easily a protein passes through the pores of the gel and size will be the major determinant of the mobility of molecules in a gel matrix. Protein shape and other factors will still affect mobility, but to a lesser extent. Substituting size for the frictional coefficient results in:

$$\textbf{Mobility} \cong \textbf{(voltage)(charge)} \div \textbf{(size)}$$

Or, in other words, the mobility of a protein during gel electrophoresis is primarily a function of its charge/mass ratio.

8.5 SDS-PAGE

Polyacrylamide gel electrophoresis (PAGE) in the presence of SDS (sodium dodecyl sulfate) is the most common form of protein gel electrophoresis. SDS completely disrupts protein-protein interactions and denatures almost all proteins resulting in a complete unfolding of proteins. In addition, ß-mercaptoethanol (or other reducing agents) is often used to break disulfide bonds. The SDS binds to the unfolded proteins giving all proteins a similar shape (i.e., random coil or extend conformation) and a uniform charge-to-mass ratio. In other words, coating proteins with a negatively charged detergent minimizes the effects of a protein's net charge. Therefore, during electrophoresis in the presence of SDS the mobility of a protein now depends primarily upon its size (i.e., mobility is inversely proportional to protein mass).

Mobility in SDS gel electrophoresis is expressed as a relative mobility (R_f). The distance the protein migrated is compared to the length of the gel, or:

$$\mathbf{R_f = distance\ protein\ migrated \div gel\ length}$$

The length of the gel is often defined by the migration of a substance which is not impeded by the matrix such a small molecular weight tracking dye (e.g., bromophenol blue). This mobility can then be used to calculate the size of proteins. Protein standards of known size are used to generate a standard curve by plotting the log of the molecular weight against the R_f values. The molecular weight of an unknown protein can be extrapolated from its R_f value. Such a calculated molecular weight is designated as Mr to indicate that it is a relative molecular weight based on comparisons to other proteins. For some proteins, though, this estimated molecular weight can differ from the actual molecular weight. In particular, highly charged proteins behave anomalously during SDS gel electrophoresis. In addition, some proteins are not completely denatured by SDS and this retention of some structure will lower the mobility.

The most commonly used procedure for checking the purity of proteins is sodium dodecyl sulfate polyacrylamide gel electrophoresis (SDS-PAGE) (see Figure 4). In electrophoresis, molecules move in an electrical field. Normally, the speed of their movement depends on three factors—their size, their shape, and their electrical charge. In SDS-PAGE, the protein mixture is treated in such a way that only the molecules' mass affects their movement. This is achieved by adding *sodium dodecyl sulfate* (C12H25- OSO3Na), the sulfuric acid ester of lauryl alcohol (dodecyl alcohol). SDS is a *detergent* with strongly amphipathic properties. It separates oligomeric proteins into their subunits and denatures them. SDS molecules bind to the unfolded peptide chains (ca. 0.4 g

SDS / g protein) and give thema strongly negative charge. To achieve complete denaturation, thiols are also added in order to cleave the disulfide bonds (**1**) (Figure 4).

Following electrophoresis, which is carried out in a vertically arranged gel of polymeric acrylamide (**2**) (Figure 4), the separated proteins are made visible by staining. In example (**3**) (Figure 4), the following were separated: **a**) a cell extract with hundreds of different proteins, **b**) a protein purified from this, and **c**) a mixture of proteins with known masses.

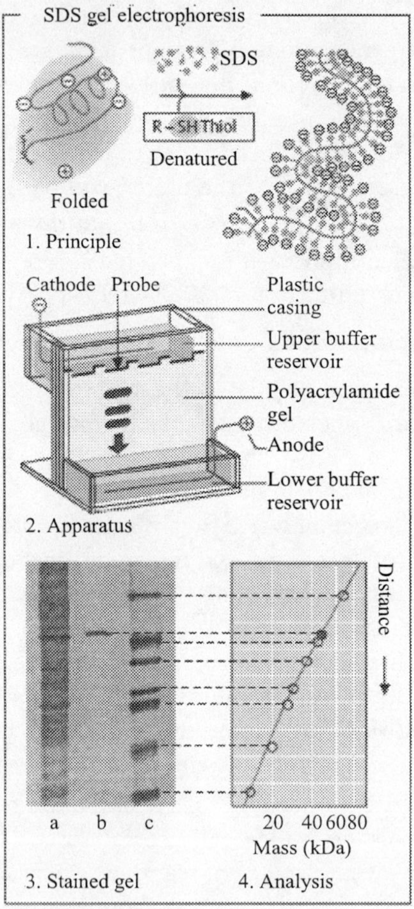

Fig.4 Descriptive flow diagram depicting sodium dodecyl sulphate-polyacrylamide gel electrophoresis (SDS-PAGE)

8.5.1 Practical Considerations

The first step in electrophoresis is to pour the separating gel (Box). Prepoured gels are also commercially available. Separating gels will typically contain 6-

20% acrylamide. The size range of the proteins being separated, the desired resolution and the amount of sample being applied are factors to consider when choosing an acrylamide concentration. Gradient gels can be used in situations where it is necessary to examine both high and low molecular weight proteins on the same gel. The stacking gel is poured after the separating gel polymerizes and just before electrophoresis to minimize diffusion between the two gels.

1. Pour separating gel.
2. Pour stacking gel.
3. Load samples.
4. Apply electric field
5. Stain or process gel.

Proteins to be analyzed by SDS-PAGE are solubilized in a sample buffer that typically contains 2% SDS and 5% β-mercaptoethanol and then boiled. The reducing agent is omitted in situations where disulfide bonds need to be preserved. In situations where an enzyme activity will be measure following electrophoresis (see below) a lower SDS concentration is used and the sample is not boiled. The amount of protein that can be loaded onto a gel is limited. Overloading the gels results in the pores becoming plugged and has an adverse effect on the electrophoresis.

After loading the samples into the wells of the gel an electric field is applied across the gel. The mobility of proteins in an electric field is proportional to field strength (E/d), or simply the voltage (E) since the distance (d) is determined by the electrophoresis apparatus. Electrophoresis will proceed faster, and therefore finish sooner, at higher voltages. However, electrophoresis generates heat in proportional to the amount of power, which is the product of voltage and current (P = EI). Excessive heating may result in proteins precipitating within the gel or have other deleterious effects on proteins depending on the nature of the protein sample and total protein concentration. These electrical parameters change during electrophoresis because the ions migrate to the anode and cathode buffers, and therefore lead to an increase in resistance (R). Resistance affects the voltage and current (E = IR) depending upon which variable is held constant. Most power supplies are capable of delivering constant voltage (E), constant current (I), or constant power (P). Electrophoresis is usually carried out under constant voltage or constant power to minimize the resulting increase in heating that occurs during electrophoresis.

A tracking dye (bromophenol blue) is included in the sample. When this dye reaches the bottom of the gel or some predetermined time afterwards the power is turned off and the proteins detected. A common way to detect proteins after electrophoresis is to stain the gel with Coomassie blue, a dye that binds

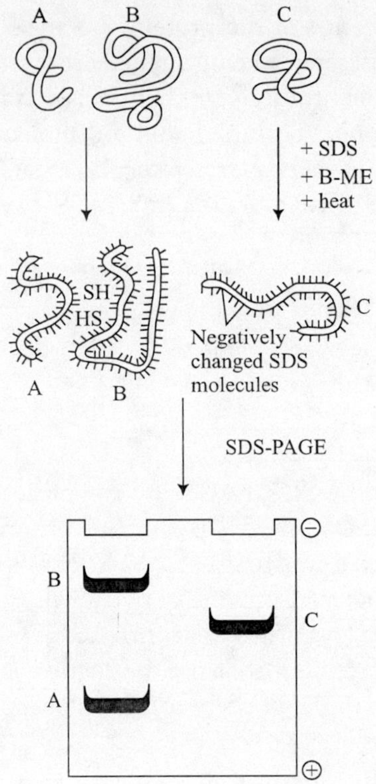

Fig. 5 Basic steps in SDS-PAGE technique applicable for different protein samples

proteins. Gels are usually 'fixed' before staining with an acetic acid and methanol solution which precipitates proteins into the acrylamide matrix.

8.6 ISOELECTRIC FOCUSING (IEF)

Isoelectric focusing (IEF) separates proteins based on their isoelectric points. The isoelectric point is defined as the pH at which a protein has no net charge (i.e., the number of negative and positive charges are equal) and is a measure of the protein's net charge. Separating proteins according to their net charge is accomplished by generating a pH gradient in an electric field. The effect of protein size on mobility is minimized by carrying out the electrophoresis gels with large pore sizes such as low acrylamide concentrations (e.g., 3.5%) or agarose. This large pore size minimizes the molecular sieving.

A pH gradient is generated with **carrier ampholytes**. These ampholytes are a mixture of aliphatic amines and either carboxylic or sulfonic acid. They have a high buffering capacity, low molecular weight (300-600 Da) and a range of pKa values. Initially the pH of an ampholyte solution will be the average of the pKa values of the mixture. Application of an electric current will cause the ampholytes to migrate toward the electrodes according to their charges. Ampholytes that have pKa values above the pH will be positively charged and those with pKa values below the pH will be negatively charged. As the ampholytes migrate, this will result in changes in the local pH due to the buffering action of the ampholytes. This change in the local pH will affect the charge on the ampholytes depending upon the pKa. The ampholytes will continue to migrate until they reach a position in which the local pH equals their pKa (i.e., no net charge) (Figure 6). The end result is a pH gradient in which the most basic ampholytes are found at the cathode, a dilute alkali solution (e.g., NaOH), and the most acidic ampholytes are at the anode, a dilute acid solution (e.g., H3PO4). Carrier ampholytes with defined pH ranges can be purchased or prepared by isoelectric focusing.

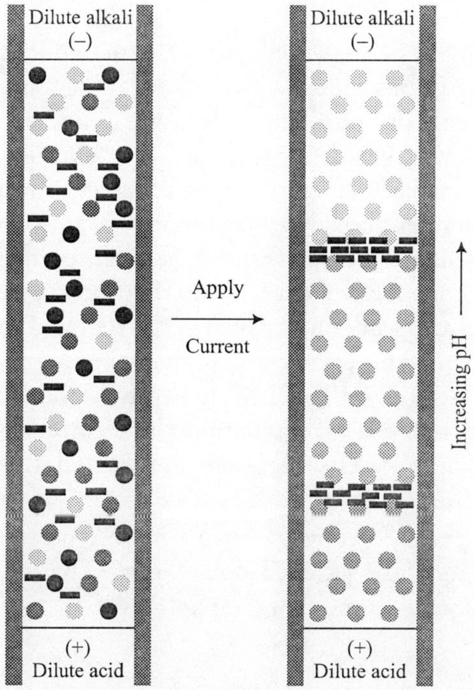

Fig. 6 Separation of ampholytes by isoelectric focusing

Proteins are also ampholytes and will migrate within the pH gradient until they reach a pH equal to their isoelectric point. The carrier ampholytes are needed since the protein concentration is generally not high enough to establish a stable pH gradient and the isoelectric points of the proteins may not be uniformly distributed along a pH gradient.

Sample preparation is important for IEF in that many reagents can adversely affect isoelectric focusing. In particular, the ionic strength should be as low as possible. Precipitation of proteins with acetone is a method for removal of excess salts, as well as concentrating the protein. Separations can be performed under either native or denaturing conditions. Urea is the preferred denaturing agent since it is uncharged. Similarly, non-ionic detergents, such as Triton X-100 or NP-40, are less likely to interfere with the formation of pH gradients. Protein precipitation is sometimes a problem in that proteins tend to be less soluble at their isoelectric points and their local concentrations can be quite high. In addition, the high voltages and high resistance (see below) associated with IEF generates substantial heat which increases protein precipitation. Many apparatuses for IEF will have a cooling mechanism to disperse the excess heat. Inclusion of urea and NP-40 in the gels will also minimize protein precipitation. In abridged form practical considerations essentially includes (i) low ionic strength, (ii) protein precipitation, (iii) heating, and (iv) gradient breakdown.

IEF is an equilibrium phenomenon since the components of the system migrate until they have no net charge. As the system approaches equilibrium the resistance approaches infinity since there are no ions to conduct the current. However, the pH gradient will start to break down before true equilibrium is reached and the ampholytes will migrate into the anode and cathode buffers. This gradient breakdown is accompanied by a lowering of the resistance. Therefore, the progress of IEF can be followed by performing the electrophoresis under constant voltage and monitoring the current. Initially the current will rapidly drop in concordance with the rapid migration of the ampholytes. As the ampholytes lose their net charge, the resistance increases and the current decreases ($E = IR$). The rate at which the current decreases levels off as the system approaches equilibrium. The current will start to rise again when the pH gradient starts to break down. IEF needs to be discontinued before this point.

The pH gradient can be determined with marker proteins with known isoelectric points or by measuring the pH along the gel. This is accomplished by slicing the gel into pieces, eluting the ampholytes into distilled water and measuring the pH.

As IEF separates proteins on the basis of charge characteristics, so this technique is utilized in the biopharmaceutical industry to determine product homogeneity. Homogeneity is best indicated by the appearance in the gel of a single protein band, exhibiting the predicted pI value. Interpretation of the meaning of multiple bands, however, is less straightforward, particularly if the protein is glycosylated (the bands can also be stained for the presence of carbohydrates). Glycoproteins varying slightly in their carbohydrate content will vary in their sialic acid content, and hence exhibit slightly different pI values. In such instances IEF analysis seeks to establish batch-to-batch consistency in terms of the banding pattern observed.

IEF also finds application in analysing the stability of biopharmaceuticals over the course of their shelf-life. Repeat analysis of samples over time will detect deamidation or other degradative processes which alter protein charge characteristics.

8.7 TWO-DIMENSIONAL GEL ELECTROPHORESIS

Conventional electrophoresis separates proteins according to their charge/mass ratios. SDS-PAGE separates proteins according to subunit, or polypeptide, mass. IEF separates proteins according to isoelectric point (or charge). It is possible to sequentially combine the different types of electrophoresis and run two-dimensional (2-D) gels. A common form of 2-D gel electrophoresis is to first separate proteins by IEF in 'tube' gels. These gels are then equilibrated in SDS gel electrophoresis sample buffer and subjected to SDS-PAGE in a 'slab' gel. (It is necessary to carry out the IEF first since SDS interferes with the isoelectric focusing.) (Figure 7). The 2-D separation results in higher resolution since proteins are being separated according to two distinct properties (i.e., charge and size).

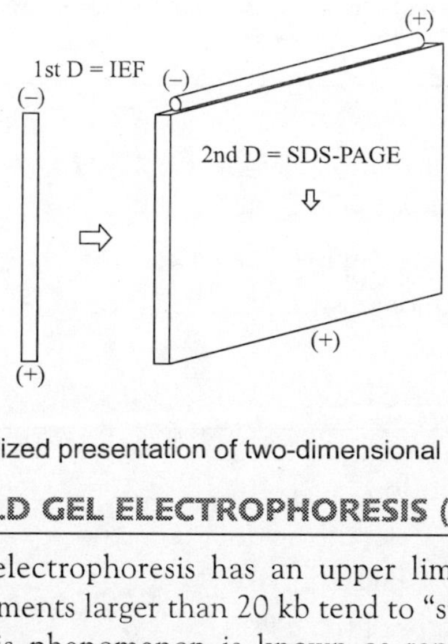

Fig. 7 Generalized presentation of two-dimensional (2-D) electrophoresis

8.8 PULSE FIELD GEL ELECTROPHORESIS (PFGE)

Conventional gel electrophoresis has an upper limit to sizes that can be resolved. DNA fragments larger than 20 kb tend to "snake" through the gel at the same rate. This phenomenon is known as reptation. Pulsed field gel electrophoresis (PFGE) overcomes the problems of resolution of large DNA fragments by using two sets of electrodes that are at fixed angles to each other. The current is alternated between these sets of electrodes at defined intervals. Separation is based upon the ability of the smaller DNA molecules to reorient faster than larger DNA molecules. DNA molecules up to 10 Mb can be resolved by PFGE.

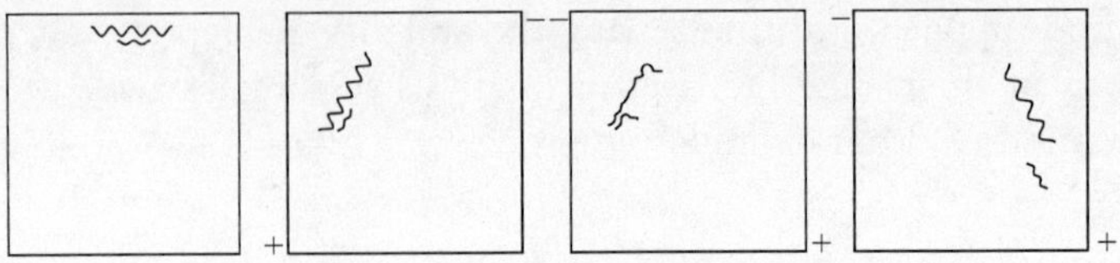

Many lower eukaryotes have chromosomes in 0.5-3 Mb size range. PFGE provides a means to characterize and map genes to the chromosomes of these organisms. In higher eukaryotes PFGE is used for long-range restriction mapping with restriction enzymes that cut infrequently.

After the initial description of PFGE in 1984, several variations of the method were developed (Box). All are based on changing the direction of migration and the differences between the methods are primarily in the positions of the electrodes. The contoured-clamp homogeneous electric field, or CHEF, is probably the more popular method. The electrodes in the CHEF apparatus are arranged in a square or hexagonal contour (see figure given below) leading to the generation of a homogeneous electric field. This results in the lanes being straight and allows for an easy comparison of different samples. In many of the other variations, the lanes run at an angle and are difficult to compare.

Variations in PFGE

- Orthagonal-field Alternation Gel Electrophoresis
- Field-Inversion Gel Electrophoresis
- Transverse Alternating Gel Electropharesis
- Programmable Autonomously Controlled Electrode
- **Contoured-Clamped Homogeneous Electric Field**

Factors Affecting Resolution

- uniformity of the two electric fields
- lengths of the electric pulses
- the ratio of the lengths of the pulses
- the angles of the two electric fields
- strengths of the electric fields

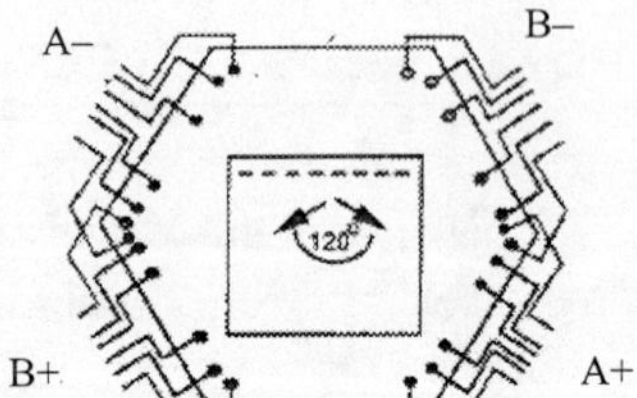

CHEF Electrodes

Large DNA fragments are easily broken by shear forces. To avoid breakage of chromosome- sized DNA fragments, whole cells are embedded into agarose blocks. The cells are lysed in the agarose block and subjected to electrophoresis. The restriction digests can also carried out in the agarose blocks before electrophoresis. Several electrophoretic variables can be manipulated and these will affect the resolution. The electrophoresis conditions will depend upon the particular apparatus being used and the size range of DNA fragments to be separated.

8.9 PROTEIN DETECTION FOLLOWING ELECTROPHORESIS

Total protein stains. Coomassie-blue staining is a popular method for the detection of proteins following electrophoresis. The procedure is to 'fix' the proteins into the gel, usually in methanol and acetic acid, and then to incubate the gel in a solution containing the dye. Excess dye is removed by destaining in the acetic acid and methanol solution. The method is cheap and easy to carry out, but of limited sensitivity.

Silver staining is generally 10-100X more sensitive that Coomassie-blue staining. Three basic types of silver staining methods have been described. All three methods are based upon binding silver ions to proteins and reducing the Ag2+ to metallic silver. The silver interacts with primary amines or sulfhydryls. In the diamine method, silver diamine complexes are stabilized with ammonium ions and the free amine groups of proteins. The silver bound to protein is then reduced to metallic silver with formaldehyde. In the non-diamine method, AgNO3 reacts with protein under acidic conditions, followed by reduction to metallic silver with formaldehyde under alkaline conditions. Photodevelopment relies on light (photons) to reduce silver ions to metallic silver. Some silver stain protocols result in certain types of proteins being stained particular colors, thus aiding in the identification of proteins.

Fluorescence. Fluorescent dyes that bind non-covalently to proteins are also used for protein detection following electrophoresis. Following incubation with the dye and destaining the gels are examined under ultraviolet illumination. Polyacrylamide exhibits some background fluorescence. Proteins can also be convalently labeled with fluorescent probes prior to electrophoresis. However, attaching fluorescent groups to proteins will change their charges.

Autoradiography and Fluorography. Radioactive proteins can be detected by exposing X-ray film to the gel. Radioactivity emitted from the proteins will activate silver grains in the film emulsion which will be converted into metallic silver during development. The reduction to metallic Ag will result in a dark spot or 'band' appearing on the film in positions which correspond to proteins. **Fluorography** refers to the exposure of film to secondary light emitted by either intensifying screens or fluors upon their exposure to radiation.

Proteins can be made radioactive by incubating cells in the presence of radioactive amino acids or other compounds that are naturally incorporated into proteins (e.g., phosphate, fatty acids, glycosyl groups, etc.). Non-physiological means can also be used to incorporate radioisotopes into proteins.

Iodination involves the reaction of radioactive iodine (^{125}I or ^{131}I) with proteins by one of several methods. The iodine in incorporated primarily into tyrosine residues and to a lesser extent into cysteine and histidine residues. Several reagents are available that will react with either free sulfhydryls (e.g., cysteine) or primary amines (e.g., lysine). Such alkylating agents can be used for the incorporation of either 3H or 14C. In most applications the proteins are radiolabeled before electrophoresis. It should be noted though that alkylating primary amines will affect the charge of the protein.

Gels are dried before carrying out autoradiography or fluorography. A gel dryer removes all of the water and solvent from the acrylamide matrix by heating the gel while under vacuum. This results in the matrix collapsing into a thin layer. The dried gel is placed next to a sheet of X-ray film in a light tight cassette and exposed for an appropriate amount of time depending on the isotope and the amount of radioactivity. The exposed film is developed.

Some emissions of high energy isotopes (e.g., ^{32}P and ^{125}I) will pass through film without exposing it. An **intensifying screen** is placed on the opposite side of the X-ray film from the dried gel. Emissions passing through the film will then interact with the intensifying screen which will fluoresce. The light from this fluorescence will then expose the X-ray film. Using an intensifying screen will usually decrease the exposure time by a factor of 7-10X. The efficiency of the intensifying screen is highest at -70°C.

Emissions from low energy isotopes (e.g., ^{3}H) and some medium energy emissions (e.g., ^{14}C and ^{35}S) are not energetic enough to escape the dried gel matrix. In this case a scintillation fluor is impregnated into the gel before drying. The radioactive emissions then interact with the fluors resulting in fluorescence. The fluorescence will then expose the X-ray film. Inclusion of fluors in the gel will not enhance the detection of high energy isotopes, and likewise, intensifying screens will not enhance the detection of low and medium level isotopes.

Phosphor-imager machines use screens containing storage-phosphors as a replacement for the use of film. These phosphor-imager screens trap the energy of radioactive emissions and are sensitive to both beta particles and gamma rays, but insensitive to visible light. The storage- phosphor captures energy with efficiency around 100% for any particle that strikes the screen. Subsequent scanning the screen with a laser beam releases the stored energy as blue light. This blue light is then converted into an image file for display and quantitation by the phosophor-imager's computer. Such high efficiency makes it possible to detect low levels of radioactivity after short exposure times. The phosphor screens can be used repeatedly by 'erasing' the screen following scanning.

Enzyme It is possible to detect enzyme activity following gel electrophoresis of some proteins. Generally it is necessary to carry out non-denaturing gel electrophoresis or IEF since proteins will need to be in a native configuration for the expression of activity. However, some enzyme activities can still be detected following SDS-PAGE. This will require that activity is located in a single polypeptide chain (i.e., no multi-subunit enzymes) and that the protein is able to refold into an active configuration following electrophoresis. To promote this refolding the sample is initially solubilized in a lower concentration (e.g., 0.1%) of SDS than normal and is not heated. Then following electrophoresis the SDS is removed from the gels by washing the gels in an appropriate buffer. The removal of the SDS and the refolding of the proteins can be facilitated in some situations by including a non-ionic detergent (e.g., Triton X-100) in the wash buffer. This will lead to a replacement of the SDS by a more appropriate detergent.

A common method to detect enzyme activity following electrophoresis is to use substrates which form insoluble products. These insoluble products will form a precipitate in the gel at the position of the enzyme activity. For example, tetrazolium salts are frequently used as color indicators for the detection of enzyme systems in which reduction equivalents are formed. The procedure is to incubate the gel with the appropriate substrates and the indicator molecules. For example, lactate dehydrogenase (Figure 8.) can be detected by incubating the gel with lactate, NAD+, phenazine methosufate (PMS) and nitro blue tetrazolium (NBT). The NADH formed by the dehydrogenease is not very efficient at transferring hydrogens and electrons to NBT and therefore an intermediate electron acceptor is needed (i.e., PMS). NBT is a soluble yellow substance which when reduced to a formazan results in the formation of an insoluble blue compound. Other dehydrogenase activities or other redox reactions are detected by using the appropriate primary substrate and changing the intermediate electron acceptor or tetrazolium salt as necessary.

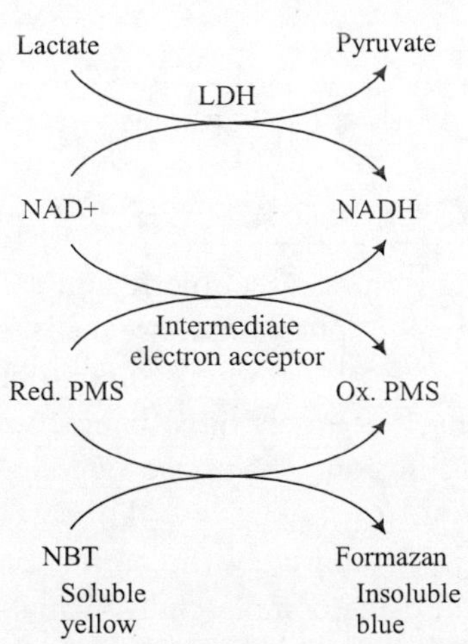

Fig. 8

It is also possible to detect enzymes which are generally not considered to be involved in redox reactions using a substrate that can be coupled to the reduction of tetrazolium salts to formazan dyes. For example, phosphatase activity can be detected by incubating with 5-bromo- 4-chloro-3-indolyl phosphate (BCIP) and NBT. After dephosphorylation of BCIP by the phosphatase, the resulting indoxyl reduces the NBT to the foramazan dye. This subtrate pair is widely used for the detection of enzyme-linked antibodies in immunoblotting procedures.

Proteases can also be detected after electrophoresis by co-polymerizing the gels with gelatin or other proteins. Following electrophoresis and removal of the SDS, the gel is incubated under conditions which promote proteolysis and then stained with Coomassie blue. Clear regions will denote the positions of proteases.

Quantification The results obtained from gel electrophoresis tend to be qualitative and difficult to precisely quantify. One possible way to quantify the amount of a specific protein is to electrophorese known amounts of a protein and to subjectively compare the staining intensities with that of the unknown protein. The stained protein bands could also by excised, the dye extracted with a solvent and the amount dye determined with a spectrophotometer. The absorbance values could then be used to generate a standard curve. However, individual proteins bind different amounts of Coomassie blue and there will be gel-to-gel variations. Similarly radioactive proteins can be excised from the gel, the gel slice solubilized and the amount of radioactivity determined by liquid scintillation counting. This method can be rather laborious if all the proteins within a sample are being evaluated.

Scanning densitometry is usually a convenient method to quantify the amount of a particular protein. Gels, auto-radiographs, or even photographs are scanned and the peak height and areas of bands are determined. The values are then used to calculate the amount of protein based on standard curves. The uniformity of the shape and width of the bands also needs to be considered.

8.10 PREPARATIVE ELECTROPHORESIS

Gel electrophoresis is a high resolution technique and can potentially be used for protein purification. One major problem, however, is the limited amount of protein that can be loaded onto a gel. Another major problem is the recovery of proteins from gels, especially if the gels have been fixed and stained. One possible method is by simple diffusion. The excised gel slice is re-suspended in a buffer and the protein allowed to diffuse out. This process is slow and the protein is excessively diluted. Alternatively, the protein can be recovered from the gel slice by electroelution using commercially available apparatuses. Both diffusion and electroelution are characterized by low yields. An alternative strategy is to transfer the protein to a membrane support and analyze the protein bound to the membrane as is done in immunoblotting or microsequencing.

If the goal is to raise antibodies against the purified protein then it may not be necessary to remove the protein from the gel. The excise gel slice containing the protein of interest is frozen in liquid nitrogen and ground into a fine powder with a mortar and pestle. The ground gel is then re-suspended in saline and used to directly immunize animals. The protein is slowly released from gel slices and actually enhances immunization through the 'depot' effect.

Commercial apparatuses for preparative electrophoresis are available. Most treat the gel as a column and continue electrophoresis until all of the proteins migrate off from the bottom of the gel. At the bottom of the gel is a chamber sealed off with a dialysis membrane to prevent the proteins from migrating into the tank buffer. The protein coming off from the bottom of the gel are continuously eluted and pumped to monitors and fraction collectors.

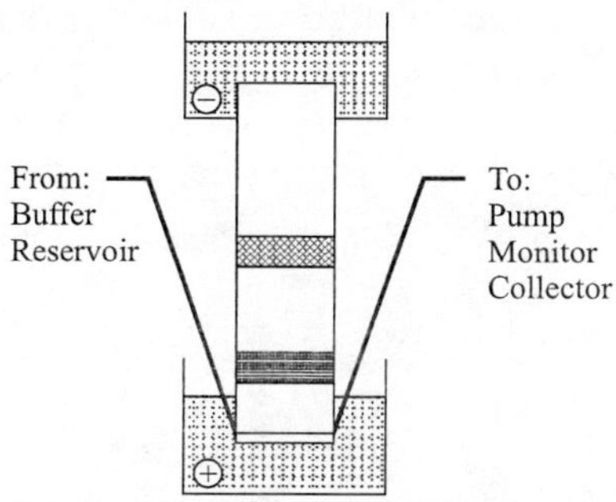

Preparative Electrophoresis

A commercially available preparative IEF apparatus, the Rotofor® (Bio-Rad), avoids the problems with gels by conducting the focusing in free solution. The apparatus consists of column (i.e., electrophoresis chamber) which is held in a horizontal position. The electrode solutions are separated from the electrophoresis chamber by ion exchange membranes and gaskets. A ceramic cooling core runs down the center of the electrophoresis chamber and the chamber is rotated around the horizontal axis during electrophoresis to dissipate heat, minimize convection, and maintain efficient electrical contact. The chamber is filled with the protein solution containing ampholytes and usually sucrose to the increase density. The mixture is subjected to electrofocusing by applying the appropriate voltage. The electrophoresis chamber is partitioned into 20 chambers each divided by a porous polyester screen with a pore size of approximately 6 μm. These screens offer some resistance to fluid convection by does not impede the flow or current or the movement of proteins. After the power is turned off the 20 fractions are

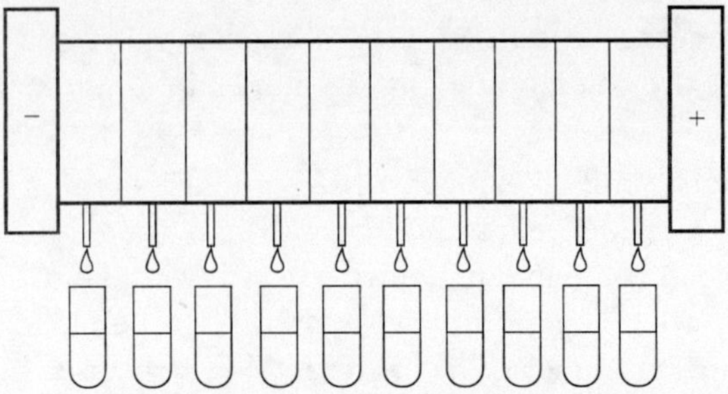

Fig. 9 Fractionation of protein solution after electrofocusing

collected with a special device that rapidly harvests the 20 fractions before substantial mixing occurs between them (Figure 9).

8.11 ELECTROPHORESIS OF NUCLEIC ACIDS

Gel electrophoresis is commonly used to separate and analyze the nucleic acids (Figure 10). The negatively charged phosphate backbone of nucleic acids gives them a uniform charge-to-mass ratio. Therefore migration in gels is inversely

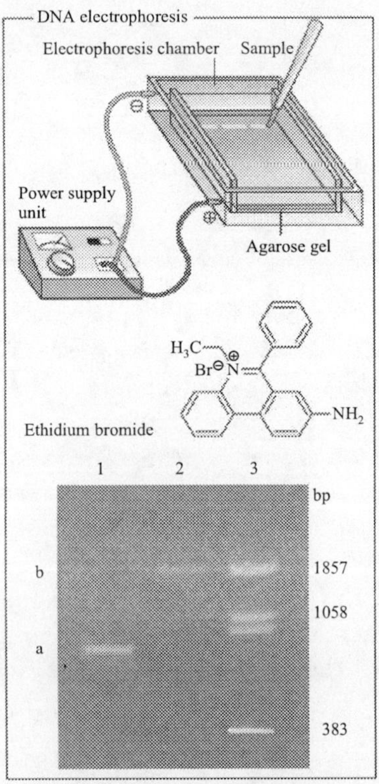

Fig. 10 DNA electrophoresis

proportional to the size of the nucleic acid (as discussed for SDS gel electrophoresis of proteins). The structure of the nucleic acid (e.g., linear, circular, double-stranded, or single-stranded) will also affect migration in gels. It is possible to distinguish DNA fragments that differ by as little as 1% and in some applications (e.g., DNA sequencing) fragments that differ in size by a single nucleotide can be resolved.

A wide range of sizes (<10 bp to >20 kb) can be analyzed by gel electrophoresis. Agarose is the most common matrix for the electrophoresis of nucleic acids. Nucleic acids ranging in size from 100 bp to 20 kb can be resolved depending on the agarose concentration . Acrylamide is used for smaller nucleic acids and oligonucleotides. Acrylamide gel electrophoresis can be carried out in the same apparatus used for protein electrophoresis (i.e., vertical). Agarose gels are usually poured and electrophoresed in a horizontal apparatus. The horizontal apparatus provides more support for the gel and are easier to prepare. The appropriate amount of agarose is dissolve in the electrophoresis buffer by heating the solution, which is then poured into a mold and allowed to cool. A well-forming comb is inserted before the agarose polymerizes. The gel is submerged in the electrophoresis tank buffer, the samples are loaded, and the appropriate voltage applied. Tracking dyes are used to monitor the electrophoresis.

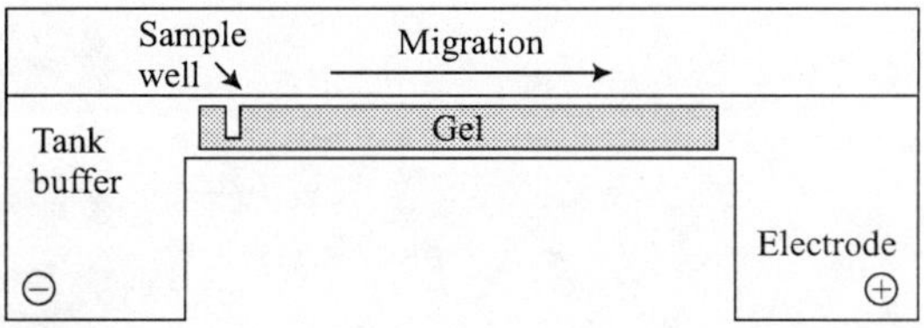

Staining with ethidium bromide (or other fluorescent dyes) is the most common method for detecting nucleic acids after electrophoresis. The ethidium bromide can by co-polymerized with the agarose gel. As the nucleic acids migrate through the gel the ethidium binds and becomes fluorescent. Illumination with UV light will reveal the fluorescent bands. The progress can be monitored during electrophoresis until the desire resolution is achieved. Alternatively the gel can be stained post-electrophoresis by incubating in an ethidium bromide solution and destaining with water. As little as 5-10 ng of DNA can be detected. Specific nucleic acids are detected by transferring the nucleic acids to a membrane and hybridizing with a DNA probe (discussed below).

Shape also effects the migration of nucleic acids in gels. For example, RNA and ssDNA can form secondary structures due to internal base pairing. These secondary structures will lead to a lower mobility. Secondary structures are eliminated by running the gels under denaturing conditions which prevent the formation of base pairs. Chaotropic agents, such as urea, formamide, or formaldehyde, are included in the gel buffers for the electrophoresis of ssDNA and RNA if sizes need to be accurately determined.

DNA can also exist as a circular molecule (e.g., plasmids). Furthermore, the circular DNA can exist in multiple conformations. In its natural state, circular DNA is supercoiled due to additional superhelical twists. Breaking a single bond in the phosphate backbone of one strand will result in a relaxed circular DNA molecule. Breaking both strands in the same position (e.g., endonuclease) will result in the formation of a linear molecule. Supercoiled circular DNA, relaxed circular DNA and linear DNA of the same molecular weight will migrate at different rates through the gel (Figure 11). The typical order of migration from fastest to slowest is supercoiled DNA, linear DNA, and relaxed circular DNA. However, the exact migration is influenced by the agarose concentration, the applied current, ionic strength and number of superhelical twists. Furthermore, circular DNA can form concatenated forms (dimers, trimers, etc.). Therefore, multiple bands are often observed in highly purified plasmid preparations. Complete digestion with a restriction endonuclease which only cuts once within the plasmid will result in the formation of a single band corresponding to the linear DNA molecule.

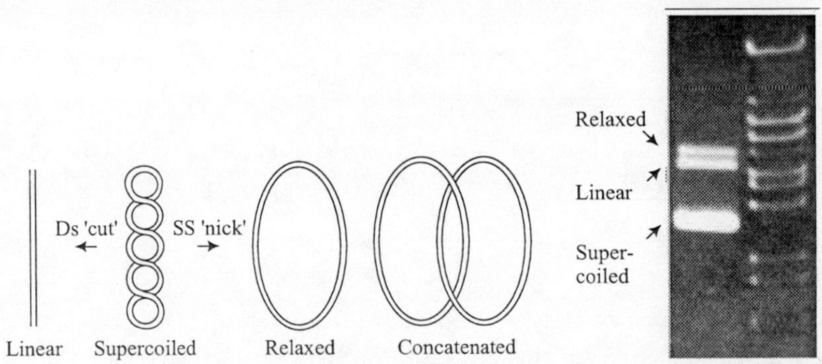

Fig.11 Band patterns of supercoiled, linear and relaxed DNA molecule(s) upon electrophoresis

Recovery of DNA from gels. DNA fragments are relatively easy to recover from gels (Box). One possibility is to transfer the nucleic acids to membranes and analyze the DNA or RNA by blotting techniques. DNA can also be recovered by electroelution. The regions containing the DNA fragment of interest is excised from the gel and put into a dialysis bag with electrophoresis buffer. The dialysis bag is placed into an electrophoresis chamber. The DNA fragment will migrate out of the agarose but remain in the dialysis bag. The DNA is then recovered by precipitation.

- transfer to membranes
- electroelution into dialysis bags
- Nal/glassmilk (eg., GeneClean®)
- low-melting temperature agarose

Another method involves dissolving the excised agarose gel piece with NaI. A silica resin which binds DNA is then added to the dissolve gel solution. Following several washes, the DNA is eluted from the resin at low ionic strength (see section on purification by absorption methods). Low-melting temperature (40-50°C) agarose is also available. The DNA fragment is excised and the gel is melted. Temperatures below 60°C will rarely have an adverse affect on the DNA. The fragment is then recovered from the solution by ethanol precipitation.

9

CHAPTER

MEMBRANE SEPARATIONS

9.1 INTRODUCTION

Membrane and membrane separation techniques have grown recently from a simple laboratory tool to an industrial process. Today membranes are used on a large scale to produce potable water from the sea by reverse osmosis, to clean industrial effluents and recover valuable constituents by electrodialysis, to fractionate macromolecular solutions in the food and drug industry by ultrafiltration, to remove urea and other toxins from the blood steam by dialysis in an artificial kidney, and to release drugs such as scopolamine, nitroglycerin, etc. at a predetermined rate for medical treatment.

In bioprocess industry, membrane separation is widely used because of the mild operating conditions and low energy requirements in the recovery of various cellular products, separation of immiscible components such as recovery of extracellular products (e.g. proteins, enzymes) and biomass form fermentation broth. Membrane separation process cannot be used for feeds containing a high concentration of low molecular weight components because of high osmotic pressure or when the feed has high solid content because of pumping problems.

Membrane processes may differ in their mode of action, in the structures used as separating barriers, and in the driving forces used for transport of different chemical components but they have several features in common which help them to be used as a separation tool. In many cases, membrane processes are faster, more efficient and more economical than the conventional separation techniques used so far. With membranes, the separation is usually performed at ambient temperature, thus allowing temperature-sensitive solutions to be treated without the constituents being damaged or chemically altered. This is important in the food and drug industry and in biotechnology where temperature-sensitive products have to be processed.

Membrane separation consists of different processes operating on a variety of physical principles and applicable to a wide range of separations of miscible components. Membrane processing is a technique that permits concentration and separation without the use of heat. Particles are separated on the basis of their molecular size and shape with the use of pressure and specially designed semi-permeable membranes. Membrane separation involves partially separating a feed containing a mixture of two or more components by use of a semipermeable barrier (the membrane) through which one or more of the species moves faster than another or other species.

Advantages of membrane separations:

- Membrane separation consistently separates a wide variety of emulsions surfactants and chelating chemistries and various mixtures.
- It requires no specific chemical knowledge.
- Complex instrumentation is not required.
- The method does not require constant attention.
- The basic concept is simple to understand.

Disadvantages:

- Membranes are expensive.
- Certain solvents can quickly and permanently destroy the membrane.
- Certain colloidal solids, especially graphite and residues from vibratory deburring operations, can permanently foul the membrane surface.
- The energy cost is higher than chemical treatment, although less than evaporation.

9.2 PRINCIPLES AND THEORY

In general sense, a synthetic membrane is a barrier which separates two phases and restricts the transport of various chemical species in a specific manner. A membrane can be homogenous and heterogeneous; symmetric or asymmetric in structure; it may be solid or liquid; it may be neutral; may carry positive or negative charges, or may be bipolar. Its thickness may vary between less that 90 nm to more than a centimeter. Mass transport through the membrane may be caused by convection or by diffusion of individual molecules, induced by electric field, or a concentration, pressure or temperature gradient.

Separation in membrane processes are the result of differences in the transport rate of chemical species through the membrane interphase. The transport rate is determined by the driving forces or forces acting on the individual components and their mobility and concentration within the interphase. The mobility is primarily determined by the solutes molecular size and physical structure of the interphase material, while the concentration of the solute in the interphase is determined by chemical compatibility of the solute and the interphase material, the solute size and the membrane structure.

The basic process of the membrane separation (Figure 1) involves a feed mixture separated into a retentate (part of the feed that does not pass through the membrane, i.e. is retained) and a permeate (part of the feed that passes through the membrane).

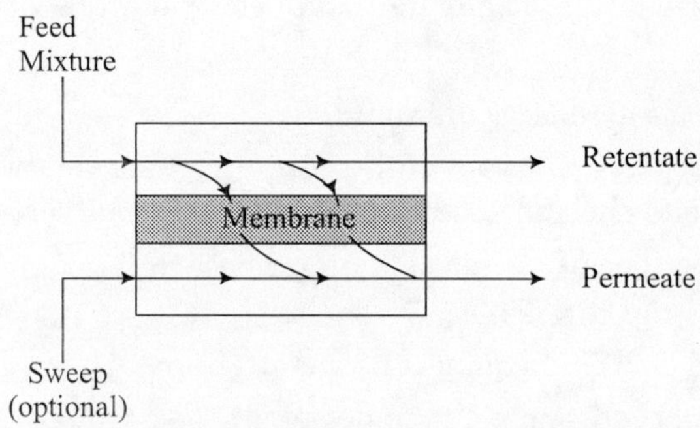

Fig. 1 A typical membrane process

When a solution and water are separated by a semi-permeable membrane, the water moves into the solution to equilibrate the system. This is known as osmotic pressure, if a mechanical force is applied to exceed the osmotic pressure (up to 700 psi), the water is forced to move down the concentration gradient i.e. from low to high concentration. Permeate designates the liquid passing through the membrane, and retentate (concentrate) designates the fraction not passing through the membrane.

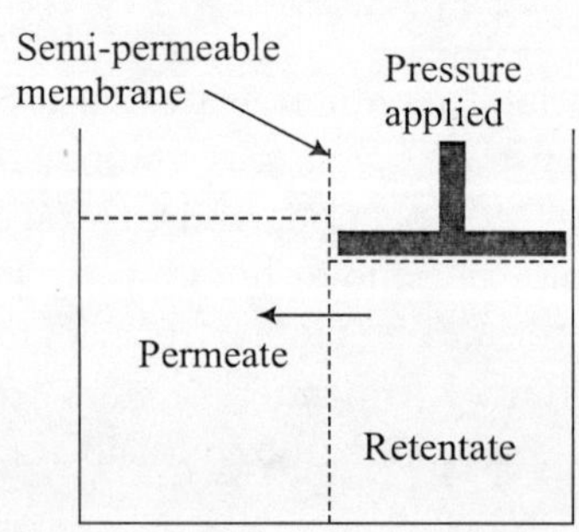

Fig. 2 Membrane processing

Permeation of liquids and gases through polymeric membranes occurs where a constituent passes through the membrane by diffusion and sorption by the fluid on the other side of the membrane. The driving force is achieved either by pressure or concentration difference across the membrane.

There are three basic forms of mass transport in membrane separation processes. The simplest form is called as "passive transport". Here, the membrane acts as a physical barrier through which all the components are transported under the driving force of a gradient in their electrochemical potential, which may be caused by differences in the hydrostatic pressure, the concentration, the temperature, or electrical potential between the two phases separated by the membrane. The second form is the "facilitated transport". Here the driving force is again the gradient in their electrochemical potential across the membrane. It is a special form of passive transport. The third form of mass transport is referred to as "active transport" in which various components may be transported against the gradient of their electrochemical potential. The driving force is by a chemical reaction within the membrane phase and is mainly found in the membranes of the living cells.

9.3 REQUIREMENT OF MEMBRANES

The operational requirements of the membranes of the pressure driven processes should have:

i. Selectivity and high separation efficiency.

ii. High permeate flux rate.

iii. Mechanical/physical strength to withstand the high pressure operation without elongation of the pores.

iv. Durability and consistency of performance over prolonged periods.

v. Resistance to corrosion.

vi. Ease of fabrication in appropriate shape.

vii. Low cost and should be readily available.

9.4 STRUCTURE OF MEMBRANES

The membrane must allow certain constituents to pass through; they must have a high permeability to certain types of molecules. In order to select the most appropriate membrane for a specific separation process, there are several factors that should be considered such as their pore size distribution, water content and sorption isotherm, fixed charge concentration, solute partition coefficients, transport parameters. Three basic types of membranes can be distinguished based on structure and separation principles:

9.4.1 Porous Membranes

Porous membranes (Fig. 3) are used in microfiltration and ultrafiltration. The dimension of the pores (0.1~10 um) mainly determines the separation characteristics. High selectivity can be obtained when the size of the solute is large relative to the pore size in the membrane. This process is basically used to concentrate high molecular weight substances and to separate high from low molecular weight substances. The interaction of molecules with the membrane matrix, facilitate the permeation of the molecule through the membrane. Microporous membranes are similar to porous membranes and differ in regards to pore.

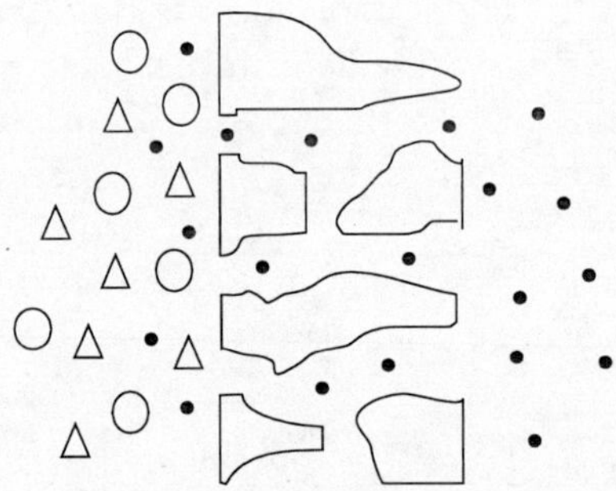

Fig. 3 Porous Membrane

9.4.2 Non-Porous Membranes

These membranes are capable of separating molecules of the same size, gases as well as liquids. Non-porous membranes do not contain any macroscopic pores. The transport is determined by the diffusion mechanism, which means that components first must dissolve into the membrane and then diffuse through the membrane due to a driving force. Also the intrinsic properties of the polymeric material determine the extent of selectivity and permeability. Separation is due to differences in diffusivity and/or solubility. These membranes can be found in gas separation, pervaporation, vapour permeation and dialysis (Fig. 4).

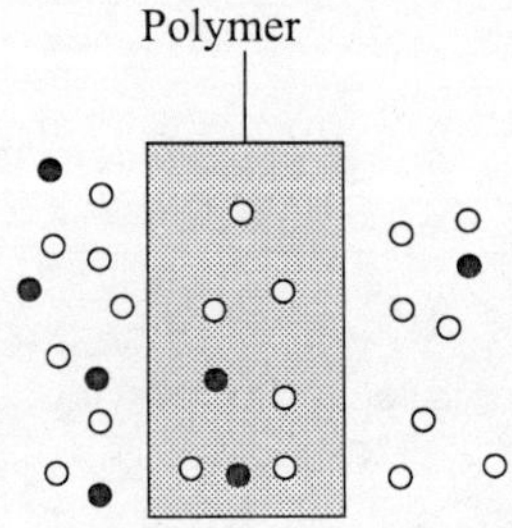

Fig. 4 Non-Porous Membrane

9.4.3 Carrier Membranes

In carrier membranes (Fig. 5), separation occurs by a carrier molecule transporting the desired component across the barrier. The carrier molecule is fixed to the membrane matrix or the carrier is mobile when it is in liquid phase. The carrier molecule shows a very specific affinity to one component or class of components in the feed, which means that high selectivity, can be obtained. Since the separation is completely determined by the carrier solute, interaction of all types of components can be removed: gaseous or liquid.

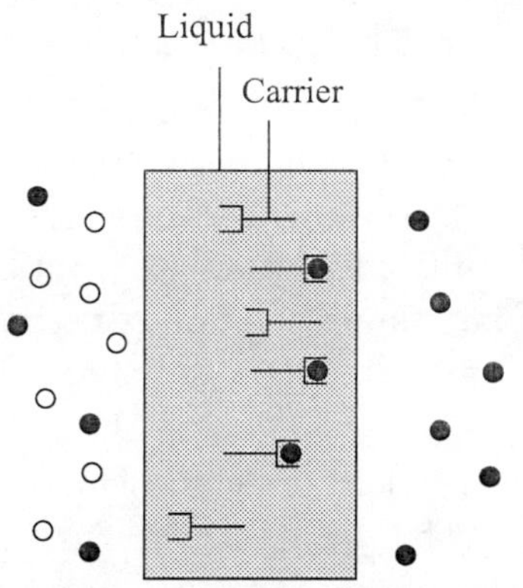

Fig. 5 Carrier Membrane

9.5 MEMBRANE TYPES AND THEIR PREPARATION

All industrial membranes are made up of natural or synthetic polymers. Natural polymers include wool, rubber and cellulose. Examples of synthetic polymers include polyamide, polystyrene etc. are most commonly used polymers and their applications are listed below in the table (see Table: 1).

Table 1 Various types of materials used for preparation of membranes used in membrane separations.

Materials	*Membrane Separation Process*
Plyroylene	Microfiltration (MF)
Polysulfone	Ultrafiltration (UF), Gas Separation (GS)
Polyimide	Gas Separation
Polyamide	Reverse Osmosis (RO)
Polyacrylonitrile	Ultrafiltration
Cellulose	Microfiltration, Ultrafiltration, Reverse Osmosis

Membranes made from natural or synthetic polymers are called as organic membranes. Apart from these membranes can also be made from other non-polymeric materials such as metals, ceramic, carbon and zeolite. Such membranes are called as inorganic membranes.

In addition to these, recent development has also introduced membranes such as liquid membranes, mixed matrix membranes and bipolar membranes. Liquid membranes consists of liquid phase that are existing in either supported or unsupported form and serves as a membrane barrier between two phases of aqueous solutions or gas mixtures. Liquid membranes are highly selective, specific and efficient but not very stable. Mixed matrix membranes are made up of molecular sieving material such as zeolite dispersed in a polymer matrix, useful mainly in gas separation. In bipolar membranes, membranes of different ionic charge are "sandwiched" together. In its simplest form a bipolar membrane is a cation-exchange membrane laminated together with an anion-exchange membrane, through an intermediate layer called as the "junction" layer.

The basic steps involved in the preparation of the membrane are:

1. Select a membrane material depending upon the separation process. For e.g. cellulose esters and nitrates can be used for the preparation of membranes for ultrafiltration and RO but cannot be used for microfiltration. A homogenous solution of desired viscosity is made by dissolving the base material in a suitable solvent with necessary additives.
2. Spread the solution on a clear, dry glass plate or draw as a hollow tube with proper thickness. This creates a thin film.
3. A controlled atmosphere is maintained to allow evaporation of the solvent from this cast film. Evaporation time and atmosphere are important factors which determine the quality of the membrane.
4. In this step gelation is done i.e. the membrane is dipped in a water bath at controlled temperature to leach out the solvent and additives slowly, thereby creating microspores in the membrane.
5. In the final step annealing is carried out by heating the membrane in water bath at 70-90°C to allow shrinkage of the pores formed.

9.6 MEMBRANE MODULES

There are several ways to economically and efficiently package membranes to provide large surface area for effective separation. Thin membranes can withstand the required operating pressure by providing them with an adequate structural support. The membranes can be cast as flat sheets, tubes and fine hollow fibers. For accommodating such shapes and structures, different types of membrane modules are available.

9.6.1 Plate and Frame

This module is mainly used in Electrodialysis and pervaporation systems and in a limited number of reverse osmosis and ultrafiltration applications. Plate and

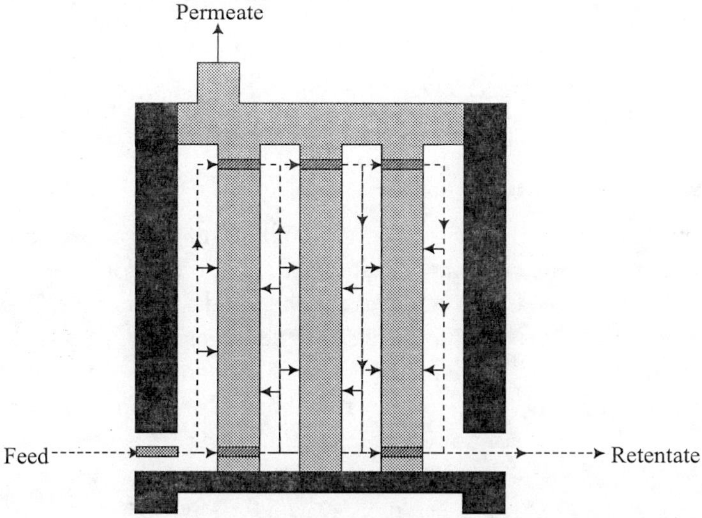

Fig. 6 Plate and Frame Module

Frame module (Fig. 6) consists of several flat sheets of membranes stacked together and forms multilayered sandwich. The module is easy to fabricate and use. Its module is compact and can withstand high pressure drop, but it has a very small membrane area per unit separator volume.

9.6.2 Spiral-wound Module

The feed passes through membrane that is spirally wound around the porous tube (see Fig. 7). The membrane, feed spacer, and permeate spacer are glued on three sides and terminates at its fourth side into the porous pipe that collects the permeate. This module is wrapped into a spiral and placed in a cylinder

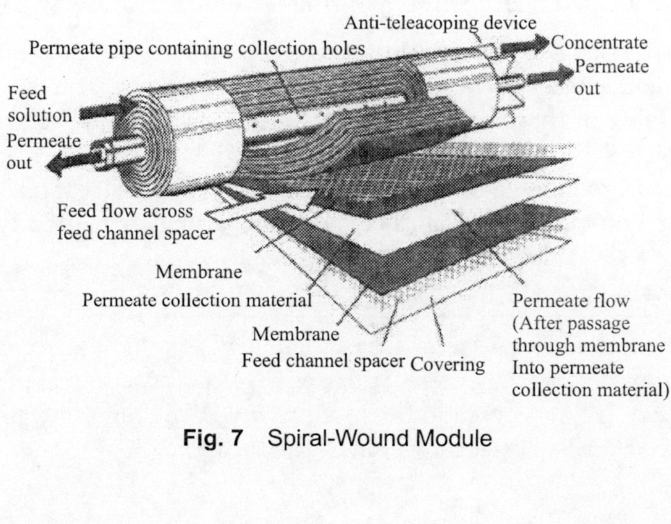

Fig. 7 Spiral-Wound Module

shell. The feed is passed axially or spirally over the outer side of the double layer. The spiral wound membrane has a high surface-to-volume ratio. The advantage of this module is that it is easy and inexpensive to adjust hydronomics by changing feed spacer thickness to overcome concentration polarization and fouling.

9.6.3 Tubular Membrane Module

Tubular membrane modules are now generally limited to ultrafiltration applications, for which the benefit of resistance to membrane fouling outweighs the high cost. This module is also called as the shell and tube module (Fig. 8) as it contains 5-7 smaller tubes nested inside a single larger tube. The feed enters the lumen of the tubes, the permeate passes through the walls while the retentate passes out at the other end of the tubes. This module is useful for highly viscous feed or feed containing particulate matter.

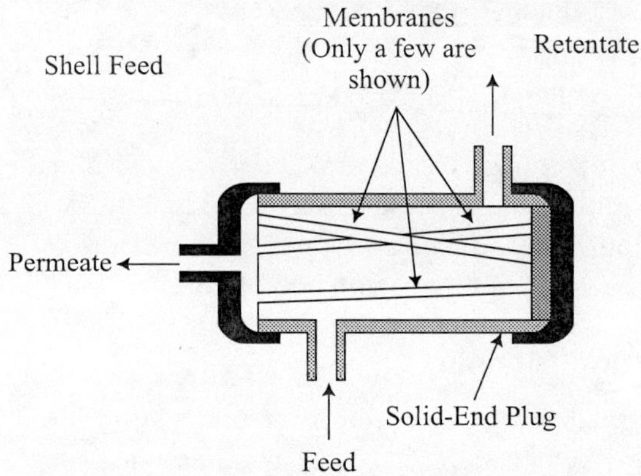

Fig. 8 Tubular membrane module or Shell & Tube Module

9.6.4 Hollow-fibre Membrane Module

This module is basically used in ultrafiltration, RO and dialysis. The application of pressure is on the inside of the fibres in ultrafiltration while it is on the shell in RO. This module consists of hollow capillary fibres of membranes packed into a shell-and-tube (Fig. 9) arrangement to provide high surface area per unit volume. The driving force is small and hence large surface area of the membrane is required. The feed must be free from particulate matter.

9.7 FLOW PATTERNS IN MEMBRANE MODULES

Membrane separation flow patterns are parallel flow, series flow and two-stage flow. Parallel flow is used when there is no need of selectivity either in permeate or retentate. Series flow is used for high selectivity in the retentate and two-stage flow is used for higher selectivity in the permeate

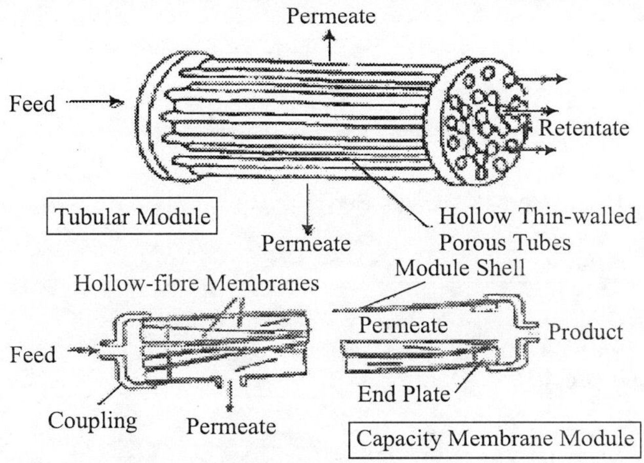

Fig. 9 Hollow-fibre membrane module

9.8 MEMBRANE FOULING

Fouling is probably the most frequently encountered problem that all membrane processes have to face. It is the coating or covering of the membrane surface of blocking of the pores which hinders the passage of permeating solution to pass through the membrane. Fouling occurs mainly due to deposition of certain material on the surface of the membrane, it can enter the module in the feed as particulate, gels or soluble, high molecular weight species or it may be due to slime formation, microbial growth, colloid deposition etc. These factors slowly decrease the flux through the membrane with time and because the flux is towards the membrane, these fouling substances tend to migrate towards the membrane surface. Fouling is irreversible and necessitates the replacement of the membrane. It can be inhibited by careful selection of membrane material, pretreatment of the feed, frequent cleaning of the membrane with chemicals and backflushing with permeate.

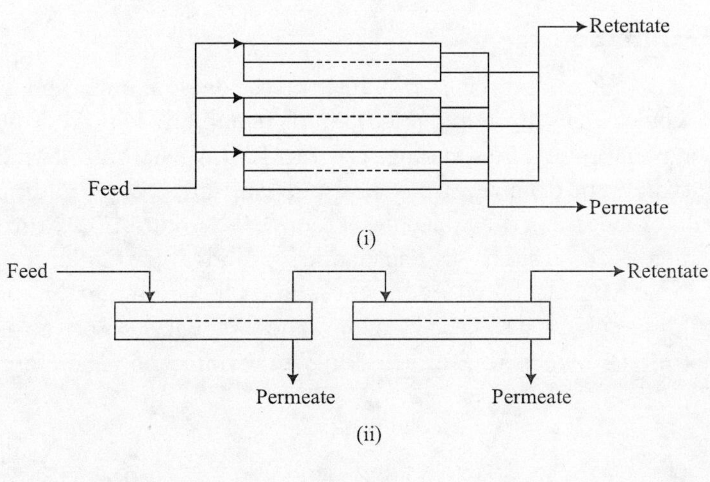

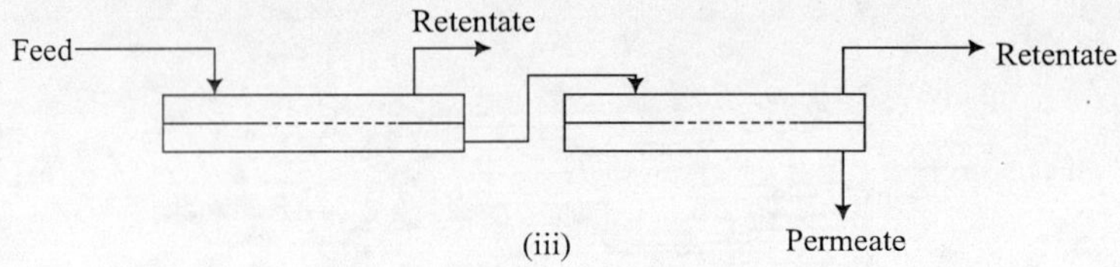

Fig.10 (i) Parallel flow, (ii) Series flow, (iii) Two-stage flow

9.9 MEMBRANE SEPARATION PROCESSES

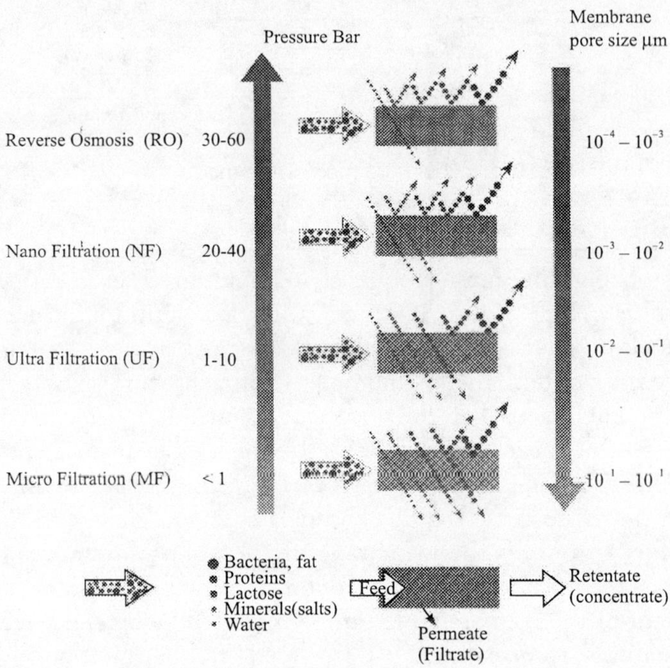

Fig. 11 Generalized description of membrane separation processes

9.9.1 Microfiltration

Microfiltration is a filtration process which removes contaminants from a fluid (liquid & gas) by passage through a microporous membrane Fig. 12. A typical microfiltration membrane pore size range is 0.1 to 10 μm. Microfiltration is not fundamentally different from reverse osmosis, ultrafiltration or nanofiltration, except in terms of the size of the molecules it retains. Microfiltration is suited to separate larger sizes, such as suspended solids, particulates, and microorganisms. This is accomplished because microfiltration membranes are thought to act like a physical sieve. The membranes are highly porous and have discernible pores even when the surface skins are asymmetric. Therefore, the

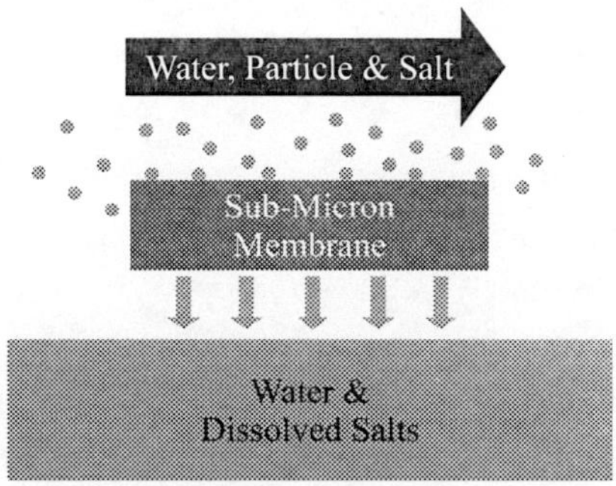

Fig. 12 Principle of Microfiltration

separation is based mainly on size. Membrane material is usually made up of ceramics, teflon, polypropylene, or other plastics.

Microfiltration is a form of filtration that has two common forms. One form is the cross flow separation Fig. 13; in this the suspension is pumped tangentially over the filtration medium. Clear liquid permeates the filtration medium and is recovered as the permeate, while the solids accumulate at the filtration barrier to form a fouling layer, or cake. The cake, constituting an increase in hydraulic resistance, decreases the permeate flux. However, the tangential suspension flow tends to limit the growth of the cake. Thus, after an initial rapid increase in cake thickness, cake growth ceases, and the cake thickness becomes limited to some steady-state value. Correspondingly, after an initial rapid decrease, the permeate flux levels off and either attains a steady-state, or exhibits a slow, long-term decline with time.

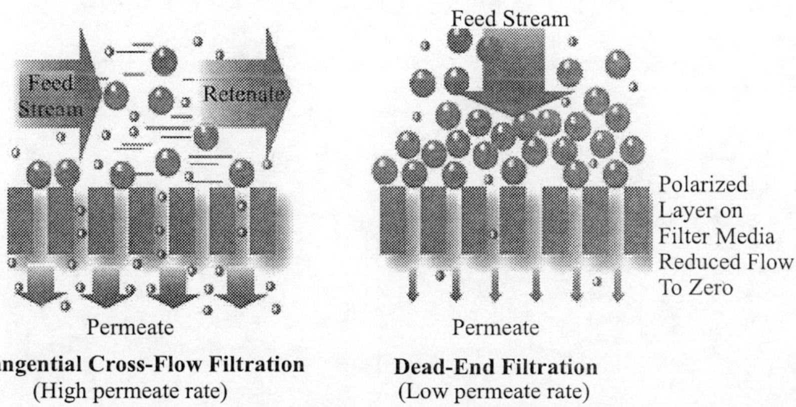

Fig. 13 Two common forms of Microfiltration

The other one is the dead-end filtration or perpendicular filtration Fig. 13. In this process, all of the fluid passes through the membrane, and all the particles that cannot fit through the pores of the membrane are stopped.

Applications:

- Fermentation industry.
- Food and beverages
- Dairy industry
- Water and Waste water treatment
- Microelectronics industry

9.9.2 Ultrafiltration

Ultrafiltration (UF) designates a membrane separation process, operated at pressure gradient of 2 to 10 bar, in which the membrane fractionates components of a liquid as a function of their solvated size and structure. The membrane configuration is usually cross-flow. In UF, the membrane pore size is larger allowing some components to pass through the pores with water. It is a separation and fractionation process. The method is useful in the separation of high molecular weight products such as polymers, proteins and colloidal materials from low molecular weight solutes.

The membrane is made up of cellulose, polysulphone or other polymers. It is finely microporous and may be asymmetric. The pore size is smaller than that in microfiltration and the pore area per unit surface is also less. The liquid flow through the membrane is by viscous flow through the pores due to the applied moderate applied pressure.

Ultrafiltration membranes are commercially fabricated in sheet, capillary and tubular forms. The liquid to be filtered is forced into the assemblage and dilute permeate passes perpendicularly through the membrane while concentrate passes out the end of the media. This may prove useful for the recovery and recycle of suspended solids and macromolecules.

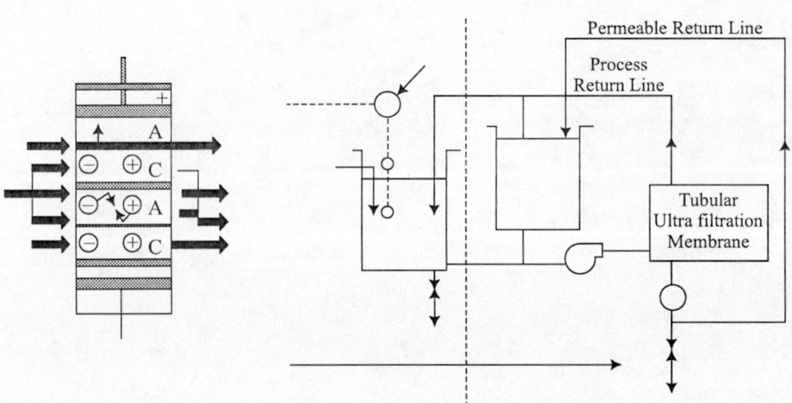

Fig. 14 Ultrafiltration

Diafiltration is a specialized type of ultrafiltration process in which the retentate is diluted with water and re-ultrafiltered, to reduce the concentration of soluble permeate components and increase further the concentration of retained components.

Applications:

- Food industry to concentrate and clarify fruit juices.
- Dairy industry - treatment of whey
- Used to concentrate cell-free fermentation broths
- Oil emulsion waste treatment
- Textile industry
- Concentration of gelatin
- Enzyme and pharmaceutical preparations

9.9.3 Reverse Osmosis

When two solutions of differing concentrations of dissolved materials are separated by a semi-permeable membrane, the liquid component will tend to flow from the lower to the higher concentrated side. In a sense, the concentration difference will tend to equilibrate across the membrane. This process is called osmosis.

If the liquid on the more concentrated side is maintained at a higher pressure, however, this process can be reversed: the solvent will flow from the concentrated side to the less concentrated side. Since the membrane blocks the passage of the dissolved waste constituents, the concentrated solution becomes even more concentrated. This process is called ***reverse osmosis***. A reverse pressure difference imposed on the membrane causes the flow of the solvent from the solution side to the pure solvent side thereby concentrating the solution. The membrane configuration is usually cross-flow. With reverse osmosis, the membrane pore size (0.0001-0.001 μm) is very small allowing only small amounts of very low molecular weight solutes to pass through the membranes.

Reverse osmosis uses membranes permeable to water but not dissolved salts even of low molecular weight. Reverse osmosis membranes are made up of polymers, cellulose acetate, polyimide and the support material is usually polysulfone.

Water filtration can only remove some suspended materials larger than 1 micron, but reverse osmosis will eliminate the dissolved solids, bacteria, viruses and other germs contained in water.

Nanofiltration is similar to reverse osmosis but the membranes are slightly more porous and can be used to separate molecules up to molecular weight of 500 Da.

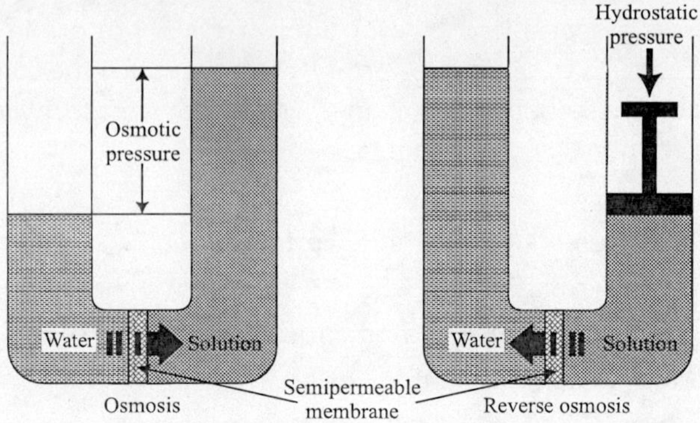

Fig. 15 Reverse Osmosis in contrast to Osmosis

Applications:

- Potable water from sea or brackish water.
- Ultrapure water for food processing.
- Pharmaceutical grade water.
- Waste treatment.
- Food and dairy industry to concentrate fruit juices, vegetable juices and milk.
- Separate low molecular weight products such as salt, sugars or organic acids from aqueous solutions.

9.9.4 Electrodialysis

Electrodialysis is used for demineralization of milk products and whey for infant formula and special dietary products. Also used for desalination of water.

Ions in aqueous solution can be separated using a direct current electrical driving force on an ion-selective membrane. Separation of ions occurs due to imposed potential differences across the ion selective cationic and anionic exchange membranes. Electrodialysis usually uses many thin compartments of solution separated by membranes that permit passage of either positive ions (cations) or negative ions (anions) and block passage of the oppositely charged ion. Cation-exchange membranes are alternatively stacked with anion-exchange membranes placed between two electrodes. The solution to be treated is circulated through the compartments and a direct current power source is applied. All cations gravitate toward the cathode (negatively terminal) and transfer through one membrane, while anions move in the opposite direction, thereby concentrating in alternative compartments.

Electrodialysis membranes are comprised of polymer chains – styrene-divinyl benzene made anionic with quaternary ammonium groups and made cationic with sulphonic groups. Anion and cation exchange membranes are arranged alternately in parallel between an anode and a cathode. The distance between the membranes is 1 mm or less. A plate and frame arrangement similar to a plate heat exchanger or a plate filter is used. The solution to be demineralized flows through gaps between the two types of membranes. Each type of membrane is permeable to only one type of ion. Thus, the anions leave the gap in the direction of the anode and cations leave in the direction of the cathode. Both are then taken up by a concentrating stream.

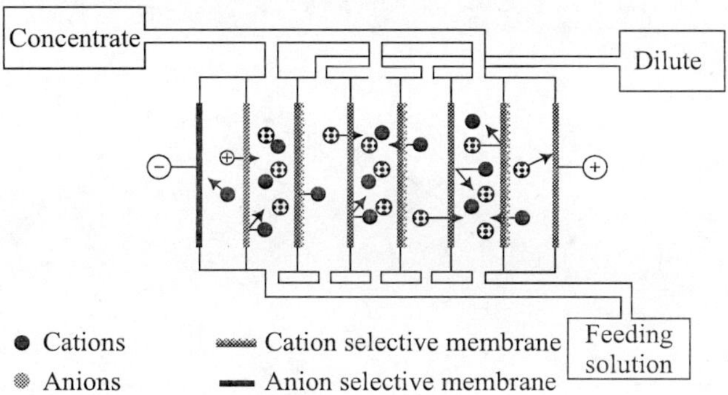

Fig. 16 Electrodialysis

Applications:

- Desalination of brackish water into potable water.
- Recover spent acid and metal salts from plating rinse.
- Removal of waste metabolites, excess water in blood.

9.9.5 Gas separation

In recent years, gas separation technology has proved to be one of the most significant unit operations. Membranes made of polymers and copolymers have been used for this process. Different gases are allowed to pass through certain membranes at significantly different rates, thus permitting partial separation. Permeation rate depends upon the differences in pressure across the membranes, thickness of the membrane, and also upon the solubility and diffusivity of the gas into the membrane.

Applications:

- Hydrogen separation and recovery.
- Syngas (synthetic gas) separation in Petrochemical industry.
- Landfill gas upgrading.
- Nitrogen production.
- Air dehydration.
- Natural gas processing.

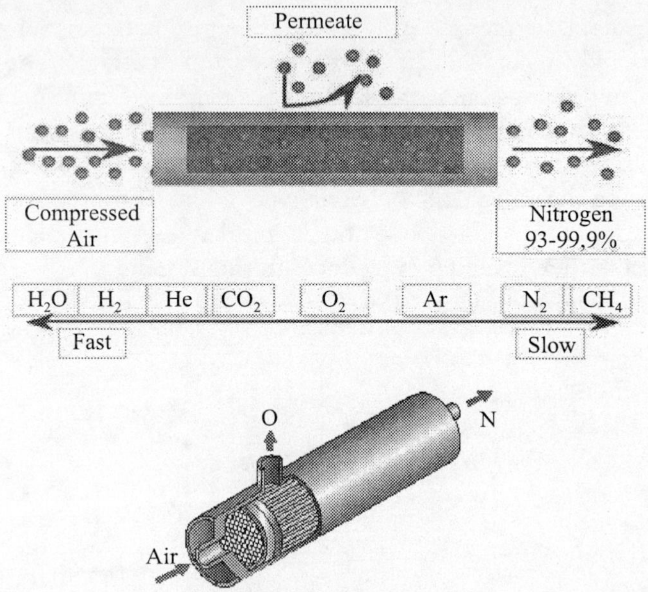

Fig. 18 Gas separation

9.9.6 Pervaporation

Pervaporation is a membrane separation technique used to separate miscible liquids, by changing the phase of the species transported across the membrane. In this process, separation is done by circulating the liquid to be separated on one side of the coated membrane and vacuum is applied on the other side. Inert carrier gas can also be used in place of vacuum, which helps in transporting the permeate away from the membrane. Separation is brought about due to the differences in solubility and diffusion. Since there is change in the phase from liquid to vapour, heat of vapourization has to be supplied to sustain the process.

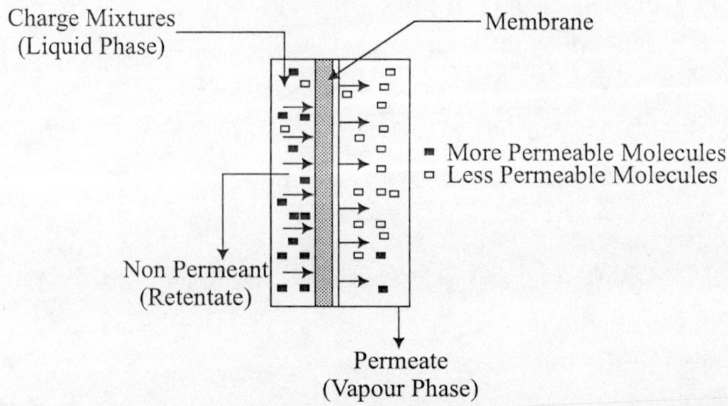

Fig. 19 Pervaporation

Applications:

- Separation of azeotropic mixtures.
- Separation of ethanol-water mixture.
- Solvent recovery.
- Enrichment of organic pollutants.
- Separation of heat sensitive products.

Summary of Membrane Separation Processes

Table 2 gives the overall summarized view for all the membrane separation processes we have discussed in prior text.

Table 2 Overall summary of various membrane separation processes

Membrane Separation	*Description*	*Initial of Feed Phase*	*Membrane Structure*	*Industrial Application*	*Driving Forces*	*Permeates*	*Concentrates*
Dialysis	Process for selective removing of low molecular weight solids from a	Liquid	Porous membrane	Hemodialysis (Removalof waste metabolites, excess body water, and restoration of electrsolution olyte balance in blood)	Presure/ Concentration Difference	Blood	Waste metabolites etc.
Gas Separation	Component(s)of a gas are removed through a pressure gradient	Vapor of Gas	Nonporous membrane	Separation of air	Pressure/ Concentration Difference	Oxygen	Nitrogen
Liquid Membrane	Extraction a solute	Vapor and/or Liquid	Liquid membrane	Recovery of zinc from wastewater	Pressure Difference	Wastewater	Zinc
Microfitration	Separation of organic and polymeric compounds with micro pore ranges of 50-500 angsstrom	Liquid of Gas	Micropourous membrane	Separaation of suspended solids from liquid mixture	Pressure Difference	Water, Solvents	Suspended solids
Pervapora-tion	Component(s) of a mixture diffuse through, evaporate undera low pressure, and are removed by a vacum	Liquid	Nonporous membrane	Dehydration of ethanol-waater azeotrope	Pressure Difference	Ethanol	Water

Table 2. Contd.

Reverse Osmosis	Passage of a solvent through a dense membrane that is pemeable to solvent but not solute(s)	Liquid	Nonporous membrane	Separation of Ions	Pressure Difference	Water	Salt Ions
Ultrafiltration	Separation of polymeric compounds from aeous solution	liquid	Microporous membrane	Separation of polymeric compounds from aqueous solution	Pressure difference	Water Low molecular weight component	Colloid, High molecular weight component

10

CHAPTER

DRYING

10.1 INTRODUCTION

Drying generally refers to the removal of moisture from a substance but more precisely it is a mass transfer process resulting in the removal of water moisture or moisture from another solvent, by evaporation from a solid, semi-solid or liquid (hereafter product) to end in a solid state. To achieve this, there must be a source of heat, and a sink of the vapor thus produced. Drying is a unit operation that is used to improve the storage life of the product and also helps in reducing the transportation cost. It can also provide definite required properties to the material as in the case of pharmaceutical substances such as tablets, syrups etc. Sometimes it can be essential part of the process such as drying of paper. Drying helps in the purification of the crystalline product so that the solvent adhering to the crystal is removed or the solvent can be recovered.

In the most common case, a gas stream, e.g., air, applies the heat by convection and carries away the vapor as humidity. Other possibilities are vacuum drying, where heat is supplied by contact conduction or radiation (or microwaves) while the produced vapor is removed by the vacuum system. Another indirect technique is drum drying, where a heated surface is used to provide the energy and aspirators draw the vapor outside the room.

In turn, the mechanical extraction of the solvent, e.g., water, by centrifugation, is not considered "drying". The ubiquitous term dehydration may mean drying of water-containing products as foods, but its meaning is more vague, as it is also applied for water removal by osmotic drive from a salt or sugar solution. In medicine, dehydration is the situation by which a person loses water by respiration, sweating and evaporation and does not incorporate, for whatever reason, the "make-up" water required to keep the normal physiological behavior of the body.

10.2 REVIEW OF DRYING OPERATION

Drying operations involve the removal of free moisture, hygroscopic moisture, or a combination of both. Free moisture is water held on the surface or between molecules of a substance. Free moisture occurs when actual liquid water is used to mix or wash the product prior to drying. Hygroscopic moisture is held within the material's cells. Hygroscopic moisture will take up or dispel water in relation to the relative humidity of the air mixture to which it is exposed. When in equilibrium with air at 100 percent rh, the material will be hygroscopically saturated. Any hygroscopic material containing free moisture must be hygroscopically saturated.

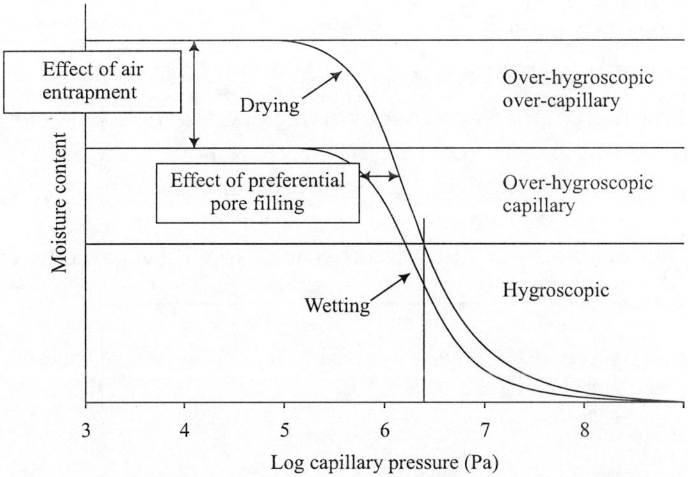

Fig. 1 Drying Operations

The removal of hygroscopic moisture depends on the relative humidity difference between that of the products' equilibrium condition and that of the surrounding air. Velocity of the air over the product has little or no bearing on the drying speed. The sudden change in drying rate (at the critical point) denotes where the initial drying via removal of free moisture ends and hygroscopic drying takes over. In other words, the product has lost its free moisture, but is still hygroscopically saturated.

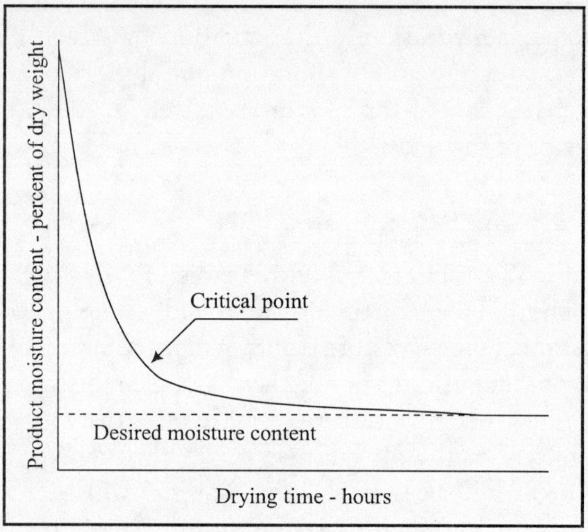

Fig. 2 Drying Curve

10.3 EQUIPMENT FOR DRYING

10.3.1 Continuous Drying

The typical flow pattern for a continuous drying operation, shown below, has a separate circulating system for free moisture removal; the dehumidifier discharge is directed through the hygroscopic moisture phase. This configuration takes advantage of rapid circulation in the first space without carrying the wetter air into the final drying space. Note the separate circulating

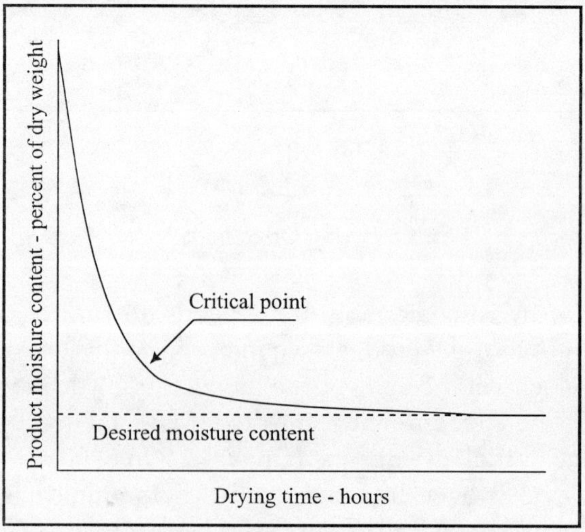

Fig. 3 Continuous Dryer

system for the free moisture removal stage dehumidifier discharge is directed through the hygroscopic moisture phase. This arrangement allows rapid circulation in the first space without carrying over moisture into the final drying space. Use the bulk drying method to establish the vapor pressure difference to allow drying to proceed satisfactorily in the free moisture stage. Keep the velocity and temperature as high as practical. Design specifications will help establish the total circulation. The temperature plus necessary vapor pressure difference will establish the specific humidity (in gr/lb.), that must be maintained.

10.3.2 Fluidised Bed Dryer

It is well known that in flash dryers, upward circulating fluidized bed dryers, bubbling fluidized bed dryers, and other particulate bed dryers, the solid's transport is affected by dispersion of the solid (annular flow or core flow) and this is a cause for different residence times and a very high dispersion in the solid's properties (final moisture, mechanical damage, thermal damage, etc.).

Compared with conventional bubbling fluidized bed or high solid concentration flash reactors (Fig. 4), the circulating fluidized bed has such advantages as high gas solid contact efficiency, high gas and solid throughput, reduced axial dispersion, and the ability to handle some particles (e.g. those that are cohesive or highly sensitive to mechanical damage) which may not be suitable for conventional beds. The lower mechanical damage to the seed, lower investment and energy consumption, and no back mixing make this operation particularly appropriate for application to the drying of biological particles. The core-annulus structure observed in co-current up-flow circulating fluidized bed and pneumatic dryers with high particle concentration induce radial solid segregation, leading not only to reduced contact between the two phases, but also to a no uniform distribution of qualities in the final product (varying solid moisture or thermal damage of seeds by an excessive exposure to high temperatures). These disadvantages of riser beds, caused by the hydrodynamic effects of both gas and solid flowing against gravity, can be largely overcome in a new type of process, the co-current down flow fluidized bed, also called the downer bed.

10.3.3 Spray Dryers

Spray drying is a commonly used method of drying a liquid feed through a hot gas. Typically, this hot gas is air but sensitive materials such as pharmaceuticals and solvents like ethanol require oxygen-free drying and nitrogen gas is used instead. The liquid feed varies depending on the material being dried and is not limited to food or pharmaceutical products. This process of drying is a one step rapid process and eliminates additional processing.

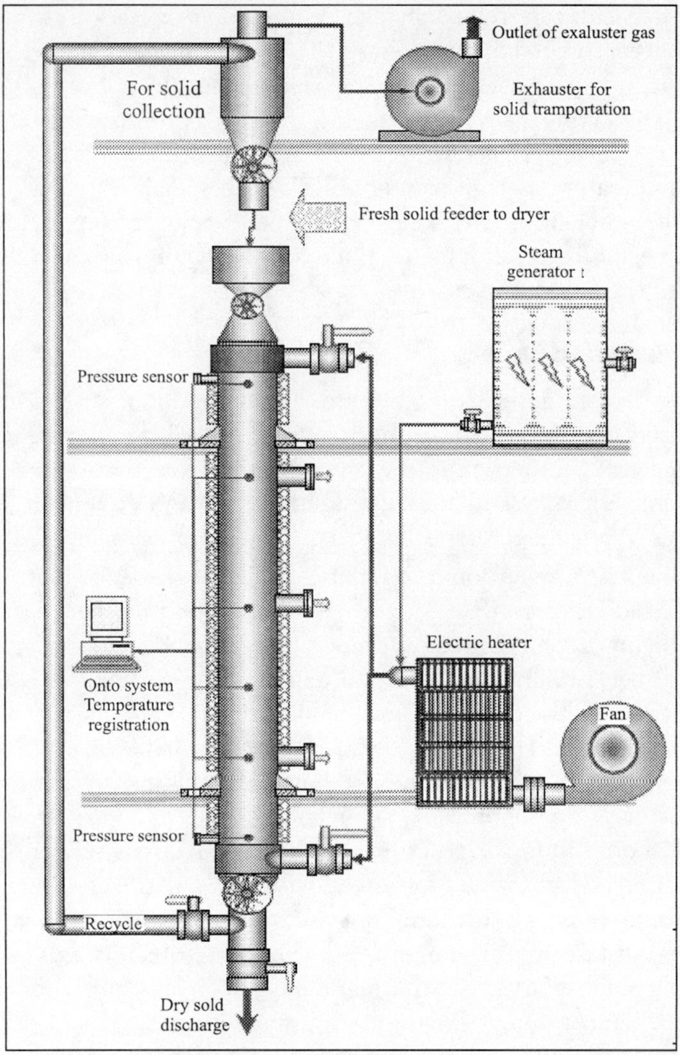

Fig. 4 Fluidised Dryer

The liquid feed is pumped through an atomiser device that produces fine droplets into the main drying chamber. Atomisers vary with rotary, single fluid, two-fluid, and ultra-sonic designs. These different styles have different advantages and disadvantages depending on the application of the spray drying required. The hot drying gas can be passed as a co-current or counter-current flow to the atomiser direction. The co-current flow enables the particles to have a lower residence time within the system and the particle separator (typically a cyclone device) operates more efficiently. The counter-current flow method enables a greater residence time of the particles in the chamber and usually is paired with a fluidised bed system.

Spray drying often is used as an encapsulation technique by the food and pharmaceutical industries. A substance to be encapsulated (the load) and an amphipathic carrier (usually some sort of modified starch) is homogenized as a suspension in water (the slurry). The slurry is then fed into a spray drier, usually a tower heated to temperatures well over the boiling point of water. As the slurry enters the tower, it is atomized. Partly because of the high surface tension of water and partly because of the hydrophobic/hydrophilic interactions between the amphipathic carrier, the water, and the load, the atomized slurry forms micelles. The small size of the drops (averaging 100 micrometers in diameter) results in a relatively large surface area which dries quickly. As the water dries, the carrier forms a hardened shell around the load. Load loss is usually a function of molecular weight. That is, lighter molecules tend to boil off in larger quantities at the processing temperatures. Loss is minimized industrially by spraying into taller towers. A larger volume of air has a lower average humidity as the process proceeds. By the osmosis principle, water will be encouraged by its difference in fugacities in the vapor and liquid phases to leave the micelles and enter the air. Therefore, the same percentage of water can be dried out of the particles at lower temperatures if larger towers are used.

The application of the spray drying encapsulation technique is to prepare "dehydrated" powders of substances which do not have any water to dehydrate. For example, instant drink mixes are spray dries of the various chemicals which make up the beverage. The technique was once used to remove water from food products; for instance, in the preparation of dehydrated milk (Fig. 5). Because the milk was not being encapsulated and because spray drying causes thermal degradation, milk dehydration and similar processes have been replaced by other dehydration techniques. Skim milk powders are still widely produced

Fig. 5 Skimmed Milk

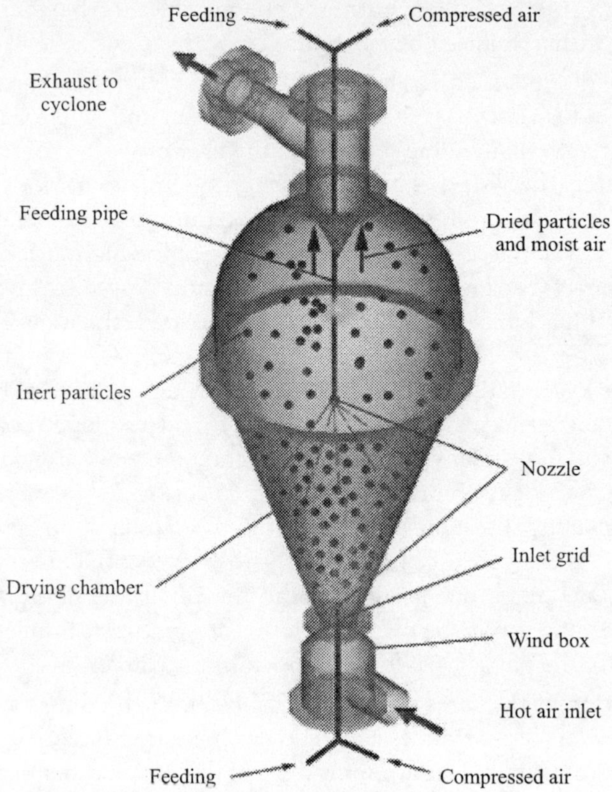

Fig. 6 Spray Dryer

using spray drying technology around the world, typically at high solids concentration for maximum drying efficiency. Thermal degradation of products can be overcome by using lower operating temperatures and larger chamber sizes for increased residence times.

10.3.4 Rotary Dryers

The **rotary dryer** is a type of industrial dryer employed to reduce or minimize the liquid moisture content of the material it is handling by bringing it into direct contact with a heated gas. The dryer is made up of a large, rotating cylindrical tube, usually supported by concrete columns or steel beams. The dryer slopes slightly so that the discharge end is lower than the material feed end in order to convey the material through the dryer under gravity. Material to be dried enters the dryer, and as the dryer rotates, the material is lifted up by a series of internal fins lining the inner wall of the dryer. When the material gets high enough to roll back off the fins, it falls back down to the bottom of the dryer, passing through the hot gas stream as it falls. This gas stream can either be moving toward the discharge end from the feed end (known as co-current flow), or toward the feed end from the discharge end (known as counter-current flow). The gas stream can be made up of a mixture of air and combustion gases from a burner, in which case the dryer is called a direct heated dryer. Alternatively, the gas stream may consist of air or another (sometimes inert) gas that is preheated. When the gas stream is preheated by some means where burner combustion gases do not enter the dryer, the dryer is known as an indirect-heated type. Often, indirect heated dryers are used when product contamination is a concern.

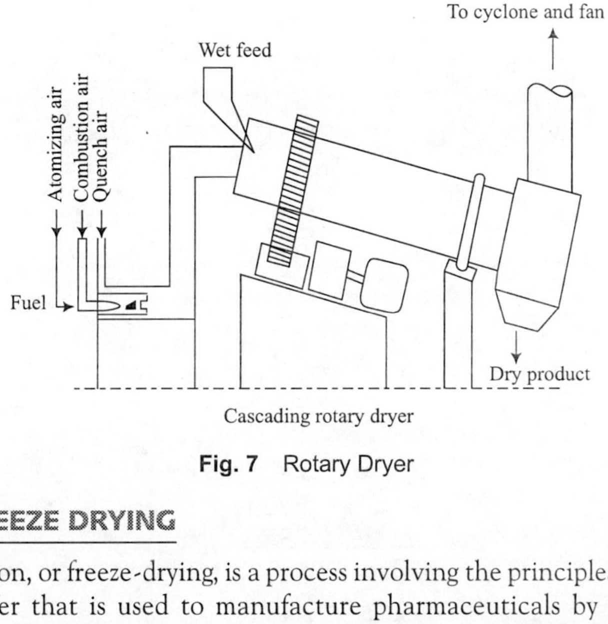

Fig. 7 Rotary Dryer

10.4 FREEZE DRYING

Lyophilization, or freeze-drying, is a process involving the principles of heat and mass transfer that is used to manufacture pharmaceuticals by freezing the product, and then allowing the water content to diminish through sublimation and desorption. This preserves the product, and is used to make parenterals that are heat-labile and capable of denaturization. It is a costly process to run; however, the products produced in a single batch may be worth millions of dollars. A lyophilization unit consists of a chamber with shelves for the products, a condenser, and a vacuum pump.

There are several advantages of using lyophilization instead of other drying processes. In the preparation of proteins and other parenteral medicines, it is important to preserve the structure and properties of the active ingredient. At high temperatures or even at room temperatures, proteins may undergo denaturation in solution. The rapid freezing during lyophilization and the removal of water allows proteins to remain in a stable state, and give a product with a longer shelf life. Lyophilization creates a porous and friable structure that can easily be reconstituted with water for injection when the drug is ready to be administered. This process also creates an accurate and sterile way to dispense the drugs into their final containers, generally glass vials.

The main disadvantage in lyophilization is the cost of equipment. It is the single most expensive unit operation in the pharmaceutical industry. This also gives rise to large energy costs. Another disadvantage of lyophilization is the long process times. The monetary gain from one batch of lyophilized product generally makes these disadvantages worth the investment.

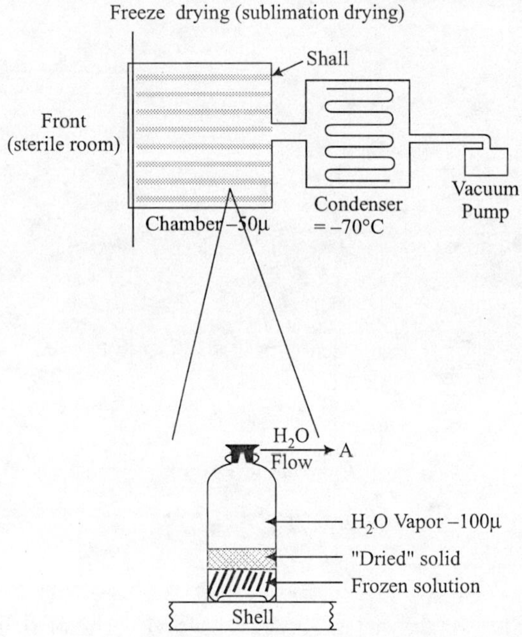

Fig. 8 Freeze Dryer

There are three stages in the whole freeze-drying process, which are (1) Freezing, (2) Primary Drying, and (3) Secondary Drying.

Freezing

The freezing process consists of freezing the material. In a lab, this is often done by placing the material in a freeze-drying flask and rotating the flask in a bath, called a shell freezer, which is cooled by mechanical refrigeration, dry ice and methanol, or liquid nitrogen. On a larger-scale, freezing is usually done using a freeze-drying machine. In this step, it is important to freeze the material at a temperature below the eutectic point of the material. Since the eutectic point occurs at the lowest temperature where the solid and liquid phase of the material can coexist, freezing the material at a temperature below this point ensures that sublimation rather than melting will occur in the following steps. Larger crystals are easier to freeze dry. To produce larger crystals the product should be frozen slowly or can be cycled up and down in temperature. This cycling process is called annealing.

Primary Drying

During the primary drying phase the pressure is lowered and enough heat is supplied to the material for the water to sublimate. The amount of heat necessary can be calculated using the sublimating molecules' latent heat of sublimation. In this initial drying phase about 98% of the water in the material is sublimated. This phase may be slow, because if too much heat is added the material's structure could be altered. In this phase, pressure is controlled through the application of partial vacuum. The vacuum speeds sublimation making it useful as a deliberate drying process. Furthermore, a cold condenser chamber and/or condenser plates provide a surface(s) for the water vapor to re-solidify on. This condenser plays no role in keeping the material frozen; rather, it prevents water vapor from reaching the vacuum pump, which could degrade the pump's performance. Condenser temperatures are typically below –50°C.

Secondary Drying

The secondary drying phase aims to sublimate the water molecules that are adsorbed during the freezing process, since the mobile water molecules were sublimated in the primary drying phase. This part of the freeze-drying process is governed by the material's adsorption isotherms. In this phase, the temperature is raised even higher than in the primary drying phase to break any physico-chemical interactions that have formed between the water molecules and the frozen material. Usually the pressure is also lowered in this stage to encourage sublimation. However, there are products that benefit from increased pressure as well.

After the freeze drying process is complete, the vacuum is usually broken with an inert gas, such as nitrogen, before the material is sealed.

10.5 DRYING IN BIOPROCESS ENGINEERING

- Foodstuffs, pharmaceuticals and biological materials which are heat sensitive even to low or moderate temperatures are freeze-dried. Freeze drying produces the highest quality product, particularly food product, with no loss of flavour and aroma and minimizes thermally induced degradation reactions.
- Spray drying is used extensively in the food and food-related industries for the manufacture of a wide range of products in dry particulate form both as powders and agglomerates. Spray drying produces these products

by atomizing a liquid formulation within a suspended particle drying system. Therefore, the spray-drying process features moist particles existing both in an airborne state and as a semi-dried product present at the walls of the drying chamber, ducts, and associated powder handling components.

- Freeze-drying is used in the preparation of proteins and other parenteral medicines as it preserves the structure and properties of the active ingredient.
- Rotary Vacuum Paddle Dryer or Paddle Dryer is used to obtain dry powder from wet cake or solutions. The drying is a batch process and often carried out under vacuum. The efficiency is very high and the operating costs are low compared to other types of dryers. The product is heated through jacket from outside and rotating hollow shaft and paddles from within. The heating is by steam or hot liquid.
- Rotary Drum Dryers are best suited for drying heat sensitive chemicals, food and pharmaceutical products. Due to short contact time with the hot drum the product can be dried at atmospheric pressure without thermal degradation and change of properties.

11

CHAPTER

CRYSTALLIZATION

11.1 WHAT IS A CRYSTAL?

Crystals are ordered structures that are built up from many billions of small identical units called unit cells. The unit cell may contain one or more than one molecule. Protein crystals are precipitated proteins in solution. They can also be called as highly concentrated aqueous solutions. Well-ordered crystals are very difficult to grow due to their irregular surfaces and it is impossible to pack them into a crystal without forming large holes or channels between the individual molecules. These holes are filled with solvent.

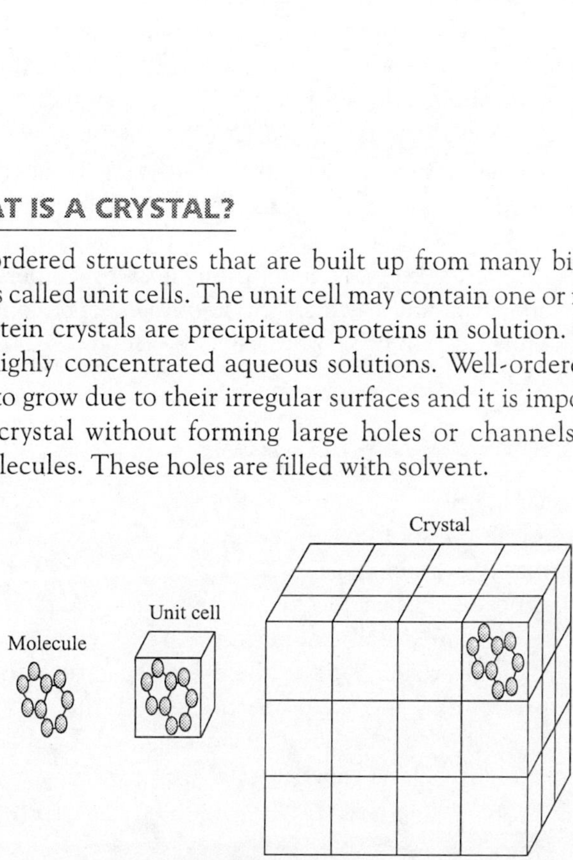

Fig. 1 Crystal

11.2 WHAT IS CRYSTALLIZATION?

Crystallization is the (natural or artificial) process of formation of solid crystals from a uniform solution. Crystallization is also a chemical solid-liquid separation technique, in which mass transfer of a solute from the liquid solution to a pure solid crystalline phase occurs. Crystallization is based on the principles of solubility, it is generally seen that compounds (solutes) tend to be more soluble in hot liquids (solvents) than they are in cold liquids. If a saturated hot solution is allowed to cool, the solute is no longer soluble in the solvent and forms crystals of pure compound. Impurities are excluded from the growing crystals and the pure solid crystals can be separated from the dissolved impurities by filtration.

Let's consider an example that is fairly easy to envision. Take a pot of boiling water and add table salt while stirring to make a water-salt solution. Continue adding salt until no more salt will dissolve in the solution (this is a saturated solution). Now add one final teaspoon of salt. The salt that will not dissolve and thus from this step the process of crystallization starts. This first step is called "nucleation" or primary nucleation. The salt resting at the bottom of the pot will provide a site for nucleation to occur. On an industrial scale, a large supersaturation driving force is necessary to initiate primary nucleation.

Crystals are grown in many shapes, which are dependent upon downstream processing or final product requirements. Crystal shapes can include cubic, tetragonal, orthorhombic, hexagonal, monoclinic, triclinic, and trigonal. In order for crystallization to take place a solution must be "supersaturated". Supersaturation refers to a state in which the liquid (solvent) contains more dissolved solids (solute) than can ordinarily be accommodated at that temperature.

11.3 PRINCIPLE OF CRYSTALLIZATION

Initial crystallization is mainly a trial-and-error procedure in which the crystals slowly precipitate from its solution by gradually increasing the level of saturation by addition of a precipitant. Precipitants remove available water forcing proteins to associate with each other. The protein precipitated out, but with appropriate care protein crystals can be grown. The precipitation point or solubility minima are usually dependent on the pH, temperature, the chemical composition of the precipitant, and the properties of both the protein and the solvent. Supersaturated solution is required for starting crystallization. The four most common methods of reaching supersaturation state are:

- Cooling (with some exceptions)
- Solvent Evaporation
- Drowning
- Chemical Reaction

Drowning describes the addition of a non-solvent to the solution which decreases the solubility of the solid. A chemical reaction can be used to alter the dissolved solid to decrease its solubility in the solvent, thus working toward supersaturation. Each method of achieving supersaturation has its own benefits. For cooling and evaporative crystallization, supersaturation can be generated near a heat transfer surface and usually at moderate rates. Drowning or reactive crystallization allows for localized, rapid crystallization where the mixing mechanism can exert significant influence on the product characteristics.

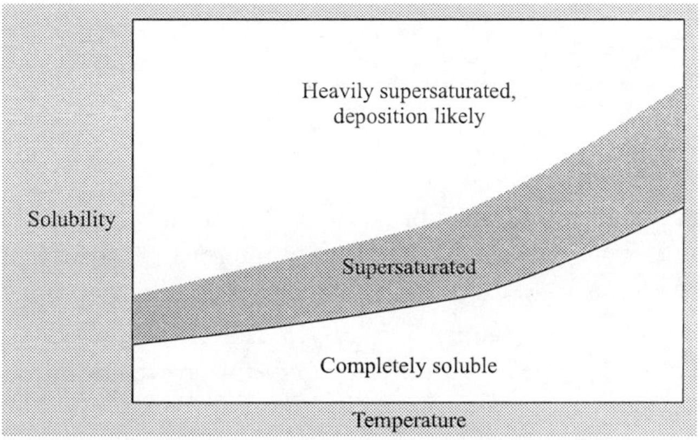

Fig. 2 Operating regions for crystallization.

A common precipitation method involves increasing the effective concentration of the protein, usually by adding a salt (salting-out) or polyethilenglycol (PEG). The salt ions and proteins compete for the solvent molecules (water) because both require hydration layers to maintain their solubility. When competition between them becomes intense, the protein molecules begin to self associate and precipitate out.

1. Feed location, undersaturated
2. Solution cools to saturation
3. Enter "metastable" zone, nucleation begins
4. Rapid nucleation
5. Concentration deereases with erystal growth
6. Crystal growth during main cooling cycle
7. Exit location, supersaturated

A large supersaturation driving force is necessary to initiate primary nucleation. The initiation of primary nucleation via this driving force is not fully understood. Usually, the instantaneous formation of many nuclei can be observed "crashing out" of the solution. The supersaturation driving force is being created by a combination of high solute concentration and rapid cooling.

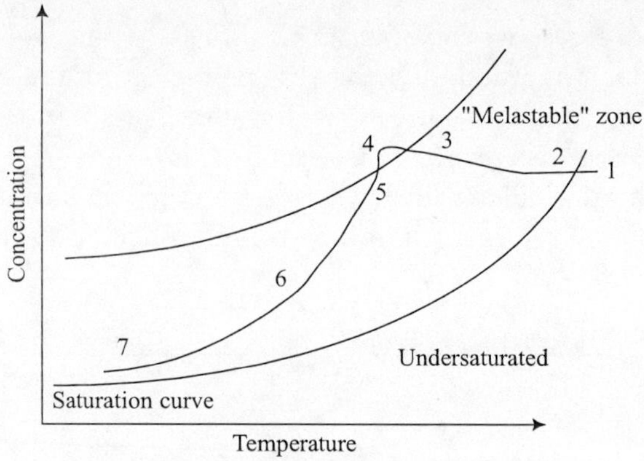

Fig. 3 Progression of Crystallization

In continuous crystallization, once primary nucleation has begun, the crystal size distribution begins to take shape. Primary nucleation is the first step in crystallization, which is simply defined as the growth of a new crystal.

The second chief mechanism in crystallization is called secondary nucleation. In this phase of crystallization, crystal growth is initiated with contact. The contact can be between the solution and other crystals, a mixer blade, a pipe, a vessel wall, etc. This phase of crystallization occurs at lower supersaturation (than primary nucleation) where crystal growth is optimal. Secondary nucleation requires "seeds" or existing crystals to perpetuate crystal growth. Secondary nucleation can be thought of as the workhorse of crystallization.

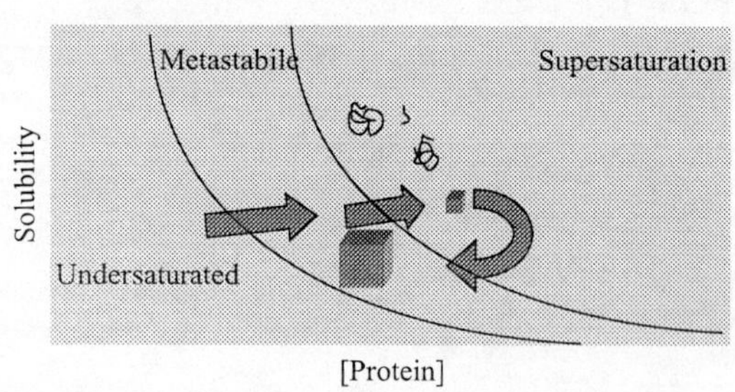

Fig. 4 Phase diagram

11.4 THEORY OF CRYSTALLIZATION

The nature of a crystallization process is governed by both thermodynamic and kinetic factors, which can make it highly variable and difficult to control. Factors such as impurity level, mixing regime, vessel design, and cooling profile can have a major impact on the size, number, and shape of crystals produced.

Now put yourself in the place of a molecule within a pure and perfect crystal, being heated by an external source. At some sharply defined temperature, a bell rings, you must leave your neighbors, and the complicated architecture of the crystal collapses to that of a liquid. Thermodynamics says that melting occurs because the entropy, S, gain in your system by spatial randomization of the molecules has overcome the enthalpy, H, loss due to breaking the crystal packing forces:

$$T(S_{liquid} - S_{solid}) > H_{liquid} - H_{solid}$$

$$G_{liquid} < G_{solid}$$

This rule suffers no exceptions when the temperature is rising. By the same token, on cooling it melts, at the very same temperature the bell should ring again, and molecules should click back into the very same crystalline form. The entropy decrease due to the ordering of molecules within the system is overcompensated by the thermal randomization of the surroundings, due to the release of the heat of fusion; the entropy of the universe increases. But liquids that behave in this way on cooling are the exception rather than the rule; in spite of the second principle of thermodynamics, crystallization usually occurs at lower temperatures (supercooling). This can only mean that a crystal is more easily destroyed than it is formed. Similarly, it is usually much easier to dissolve a perfect crystal in a solvent than to grow again a good crystal from the resulting solution. The nucleation and growth of a crystal are under kinetic, rather than thermodynamic, control.

11.5 EQUIPMENT INVOLVED IN CRYSTALLIZATION

11.5.1 Tank Crystallizers

This is probably the oldest and most basic method of crystallization. In fact, the "pot of salt water" is a good example of tank crystallization. Hot, saturated solutions are allowed to cool in open tanks. After crystallization, the mother liquor is drained and the crystals are collected. Controlling nucleation and the size of the crystals is difficult. The crystallization is essentially just "allowed to happen". Heat transfer coils and agitation can be used. Labor costs are high, thus this type of crystallization is typically used only in the fine chemical or pharmaceutical industries where the product value and preservation can justify the high operating costs.

11.5.2 Scraped Surface Crystallizers

One type of scraped surface crystallizer is the Swenson-Walker crystallizer, which consists of an open trough 0.6 m wide with a semicircular bottom having a cooling jacket outside. A slow-speed spiral agitator rotates and suspends the growing crystals on turning. The blades pass close to the wall and break off any deposits of crystals on the cooled wall. The product generally has a somewhat wide crystal-size distribution.

11.5.3 Forced Circulating Liquid Evaporator-crystallizer

These crystallizers combine crystallization and evaporation, thus the driving forces toward supersaturation. The circulating liquid is forced through the tube side of a steam heater. The heated liquid flows into the vapor space of the crystallization vessel. Here, flash evaporation occurs, reducing the amount of solvent in the solution (increasing solute concentration), thus driving the mother liquor towards supersaturation. The supersaturated liquor flows down through a tube, then up through a fluidized area of crystals and liquor where crystallization takes place via secondary nucleation. Larger product crystals are withdrawn while the liquor is recycled, mixed with the feed, and reheated

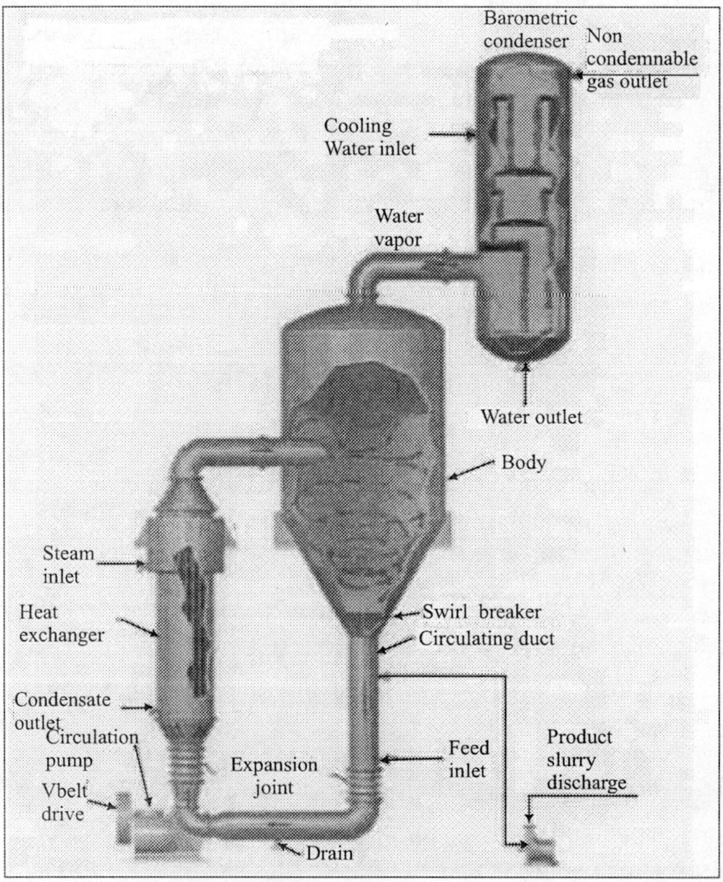

Fig. 6 Forced Circulating Liquid Evaporator-Crystallizer

11.5.4 Circulating Magma Vacuum Crystallizer

In this type of crystallizer, the crystal/solution mixture (magma) is circulated out of the vessel body. The magma is heated gently and mixed back into the vessel. A vacuum in the vapor space causes boiling at the surface of the liquid. The evaporation causes crystallization and the crystals are drawn off near the bottom of the vessel body.

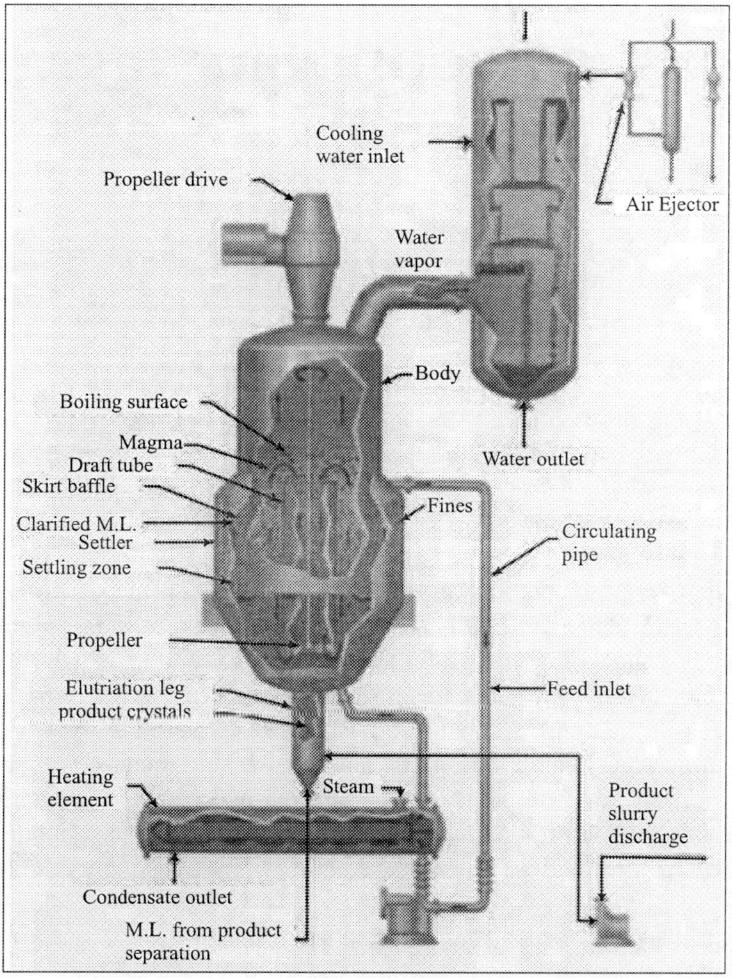

Fig. 7 Circulating Magma Vacuum Crystallizer

12

CHAPTER

FINISHED GOOD FORMULATION AND PACKAGING

12.1 INTRODUCTION

The production of biologicals involves processes and materials, which themselves have inherently high variability and show a sensitive response to small changes in physical and chemical parameters. So far, to guarantee (or to try to guarantee) consistent product quality, the principle aim has been to control, i.e., repeat, a process very stringently, but the impact of many factors along the multi-step process line and their mutual interconnectivity makes it difficult to have low batch-to-batch variation. Therefore, better understanding of the bioprocess, and better *in situ* monitoring tools are required, which are pre-requisites for feed-back control and consistent quality. There are specific needs for rapid measurement in cell banking (e.g. genotypic studies, phenotypic characterization, safety detection), preparation phases (e.g. media composition, contamination), during fermentation (e.g. cell density, cellular activity, contamination control, product formation kinetics) and DSP (e.g. impurities, contamination, intermediate product stability, product quality, column lifetime).

The global pharmaceutical market is currently estimated at about $500 billion. In 2003, almost 50 percent of the global market value was generated in North America, whilst about 25 percent was generated in the EU. Japan was the third largest pharmaceutical market generating about 10 percent of the global market value. In 2004, the 10 biggest pharmaceutical companies were estimated to generate 41 percent of the total market value. In the US, the pharmaceutical market value growth is expected to remain mediocre through 2007.

It is necessary to develop various tools, which are specific to certain process phase, production organism, or end product, but there is a more general demand for tools to control and optimize bioprocess manufacturing; fermentation models and simulation tools in combination with metabolic flux analysis, at- or on-line analytical measurements, control strategies for dynamic bioprocesses, and supervisory software tools.

From a company activity perspective, the pharmaceutical industry can be divided into the following four general phases:

1. *R&D* (target identification, discovery of active pharmaceutical ingredients (API's), pre-clinical and clinical).
2. *Manufacture of the API* for either clinical trials or for the commercial final drug product.
3. *Manufacture of the final drug product* (hereafter called Formulation).
4. *Marketing & sales* of the drug.

12.2 REVIEW OF FORMULATION OPERATIONS

Upon a fermentation process, it is apparent that most recovery processes involve combinations of the following procedures:

- Fermentor harvest.
- Mechanical separations of cells from fermentation broth (clarification).
- Optional disruption of cells.
- Extraction of the compound of interest.
- Preliminary fractionation procedures.
- High-resolution separation steps.
- Concentration.
- Formulation.
- Optional drying.

In contrast to fermentation processes, where typically one format (i.e., a fermentor vessel), is used, a number of widely differing techniques are necessary to accomplish purification and formulation of biological products. The variety of available separation technologies is large and the order and permutations of sequential processing steps are countless. Although this may hamper standardization of downstream processes, it also provides the benefit of flexibility because different unit operations can be tailored to the greatest efficiency in a cost-effective manner. Possibly a better way to illustrate a downstream process design is shown below, in Figure 12.1.

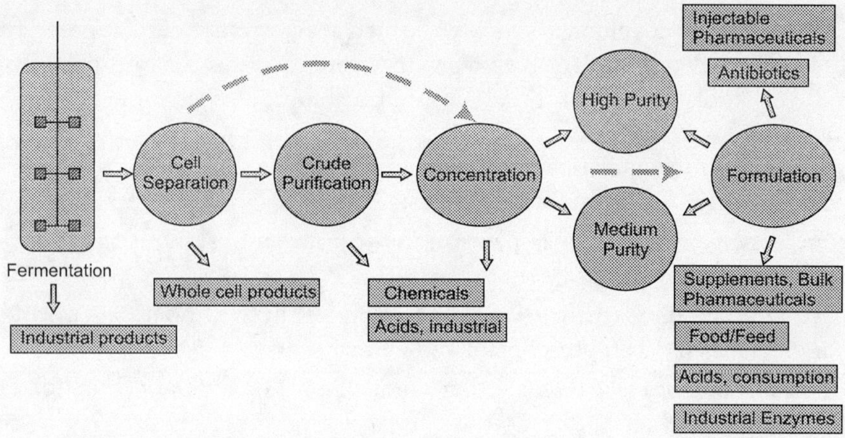

Fig. 12.1 Downstream processing cascade arranged by final product requirements

Here it is demonstrated that recovery is not a linear sequential process but rather the downstream process development follows a palette of options that can be applied depending on the required specifications of the end product and as such, it is a modular approach. Several general considerations limit the range of practical choices for a process design. The type of production organism (bacteria versus fungi or mammalian cells) has an impact on the downstream process design. The fermentation media for these organisms can differ substantially and the media themselves can play an important role for the recovery strategy. Defined media often deliver a purer product but lower in concentration, whereas complex media contain more impurities but also generate more amount of product. The physical, chemical, and biochemical properties of the desired product have an impact on the recovery strategy, as do the microbial cell's physiological states and properties. Proteins and enzymes require more "gentle" conditions to avoid unfolding or chemical denaturation than those required for small molecules. General parameters to observe are large swings in or extreme changes in pH values and temperatures. Both proteins and small molecules can have hydrophobic, hydrophilic, or limited solubility properties that need to be considered when devising a recovery and purification strategy.

As can be seen from Figure 12.1, *the nature of its final application* will dictate how pure the product needs to be and which raw materials may be used in its manufacture. Typically, high purity is not required of industrial enzymes. In contrast, therapeutic proteins and small drug molecules require the highest level of purification. The purity range for food and feed compounds is somewhere in the middle. Raw materials must meet regulatory requirements for the application and must often be tested for their toxicological properties.

Finally, economic considerations constrain both the range of raw materials and separations processes that can be used. Although multiple chromatographic steps are routinely included in the downstream processing of human therapeutic proteins, few industrial proteins include even a single chromatographic step. Purification, if any, of industrial products is accomplished by less expensive techniques such as extraction or crystallization.

12.3 FORMULATION

Formulation of a product depends on how the end product will be used and the ability to satisfy safety requirements. Many bioproducts are supplied in liquid form, as it is convenient to meter and use. If the compound is slated for dry products, the concentrate can either be stored for later use or applied directly for granulation. In many industries, proteins, peptides, and other bioproducts are supplied as encapsulates or granules in order to meet safety standards.

The formulation methods utilized depend foremost on the application of the final product. Other influencing factors include, for example, the purification and stability requirements, as well as API stability (temperature, pressure, etc.). Generally, the final formulation is either liquid (concentrates) or solid (powder, tablets, etc.).

As with fermentation and downstream processing, precise formulation recipes are challenges faced by the formulator of biological products often kept as closely guarded trade secrets. Comparatively few publications on new approaches or new technologies are available in the literature. However, the intellectual property landscape is filled with composition of matter patents.

Proteins are large molecules comprised of tens to hundreds of amino acid subunits. Their unique properties are associated with specific conformations of the amino acid chain. The active conformations are thermodynamically quite fragile, typically only a 5–15 kcal/mol difference in free energy between the stable and the unfolded conformations. Additionally, the amino acid subunits that define the protein contain a variety of reactive groups that are subject to chemical degradation pathways resulting in the loss of structure and/or activity. Thus, ensuring the physical and chemical integrity of a protein's native structure is a primary concern when using proteins or enzymes in applications where its structure is important. Many proteins are only soluble up to 20–30% (w/w) and therefore considerations for keeping the compounds well solubilized are of importance.

12.3.1 Dry Formulations

Because of the long term instability of proteins in aqueous solution, enzyme producers and formulators have attempted to produce stable solid formulations since enzymes were first used. Proteases tend to degrade via autolysis and can be incompatible with surfactants. These problems are easily overcome by storing the enzyme in the solid state. Initially, commercially produced enzymes for laundry detergents were spray-dried or sprayed onto salt cores without coating. Spray-drying provides a fast and cost-effective way of compartmentalizing two incompatible ingredients and maintaining enzyme activity over a product's typical shelf life. However, non-coated granules have led to cases of allergic reactions in formulating plant personnel and resulted in a setback for the enzyme industry in 1970s. Thus, the technologies used to produce modern dry products have evolved far beyond simple spray-drying.

Product safety is an important consideration as many bioproducts can cause adverse responses in humans and many therapeutic fermentation products are very potent bioactive molecules. Negative effects can include irritation and allergic reactions, which are due to the possible immunogenic nature of many biological compounds. Lyophilization, or freeze-drying, which is used in therapeutic applications results in a powdered product that is contained in vials as an additional safety measure to prevent direct exposure. It offers a gentle way of protecting a compound's fragile nature because the drying temperature is kept below freezing. However, these technologies are expensive.

Many of today's bioproducts are supplied in a granular form. Granulation is a generic term for particle size enlargement. Granules in consumer products need to provide a tough barrier to prevent release of the bioactive molecules in airborne dust, while providing quick release once used in the final application. Enzyme granules contain coatings that are designed to withstand the physical impact and shear forces typically encountered during powder processing, maintaining excellent flow ability throughout the process.

Many ways of producing enzyme granules have evolved, with just a handful of methods being currently in common use. These include prilling (spray-chilling), marumerization/ spheronization, drum and high shear granulation, and fluid-bed coating. The latter two techniques are superior in producing low dust granules. In general, fluid-bed technology (Figure 12.2) is the most flexible approach, giving granules with the most uniform appearance and smooth coatings. Other technologies may have an edge in either cost or throughput, as they are more amenable to continuous operation. The advantage of the fluidized-bed method is that the entire granulation process can be carried out within a single, contained piece of equipment. The spray-coating process allows the sequential application of layers of different thicknesses and compositions with almost infinite flexibility.

Solid formulations can provide some significant advantages, such as enhanced stability, delayed or controlled release, and protection against deactivation during harsh applications. One example of the latter is the use of

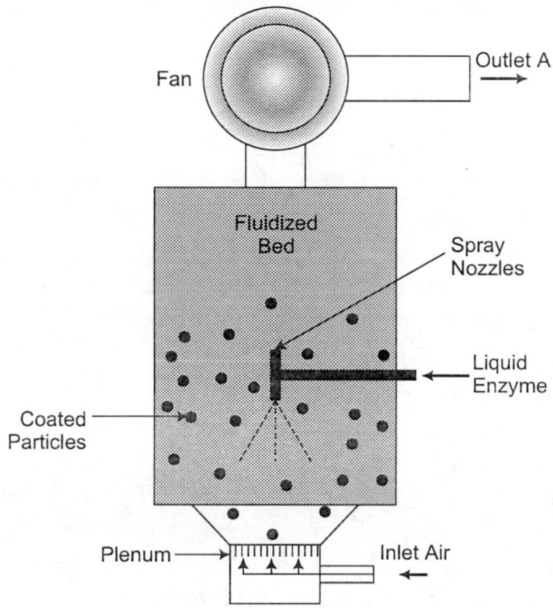

Fig. 12.2 Schematic of a fluidized bed coater

granules to encapsulate cellulase enzymes against deactivation in the steam-pelleting process that is used in producing animal feed pellets. The use of effective stabilizers and coatings can prevent or retard exposure to moisture under high heat, yet allow release of the enzyme in the subsequent feed application.

Many therapeutic molecules can be encapsulated using other techniques, such as coacervation, micro- or nanospheres, microcapsules, or liposomes. Coacervation, often used for flavor encapsulation, can be a simple aqueous phase separation of immiscible droplets, or in the case of complex coacervation, the use of opposite electrostatic charges. It is one of the basic processes of capsule wall formation. Other biochemicals, such as flavors, vitamins, or citric acid are sold directly in the crystallized forms and can be incorporated directly into the final product, such as tablets, capsules, granules, and so on.

12.3.2 Liquid Formulations

The successful application of enzymes in liquid formulations presents several technical obstacles that are not encountered in powders and granules. These problems stem from the fact that liquid products are complex aqueous solutions, and physical separation of enzymes from other potentially inactivating ingredients is impractical. Necessary ingredients for a viable end-product can affect the physical and chemical stability of enzymes. In addition, as newer types of enzymes are added to formulations that already contain proteases, proteolytic degradation of enzymes is a concern.

Liquid formulations can be used as pastes or slurries, dispersions, or emulsions (in aqueous or other solvents) to physically separate the compound to be protected. Dispersions, pastes, or slurries are not commonly used but are a feasible approach when highly concentrated products are needed. For instance, many early analytical enzymes were sold as crystal slurries to preserve stability. Dispersions can be prepared from dry materials. Most of the dispersions are visually cloudy in appearance due to the particle size of the dispersed compounds and are thus undesirable in most end-products.

Many catalytically active proteins are hydrolases and are subject to three principal means of deactivation: denaturing or unfolding, catalytic site inactivation, and proteolysis. Denaturation is best minimized by controlling temperature and pH and by avoiding the presence of chemical denaturants. Catalytic site inactivation is prevented by supplying sufficient levels of cofactor, typically a metal cation, and preventing oxidation of the active site, for example, by formulating with antioxidants. Alternatively, oxidative resistance can be engineered into the protein structure. Finally, in the case of proteases, proteolysis or autodigestion can be minimized by reducing water activity, or by addition of inhibitors. Thus, the final liquid formulation includes stabilizers, antimicrobial substances, and osmolytes to reduce the water activity and increase thermodynamic stability. Useful water-sequestering compounds include sugars and other polyols, such as glycerol, sorbitol, and propylene glycol. Useful inhibitors include substrate analogues such as peptides and acid salts. Once the formulation excipients are added, the last step includes a final polish filtration to provide a clear liquid.

Different industries pose different challenges to the protein formulator. Many feed enzymes are sold as formulated liquid concentrates. In this case, the major requirements for a liquid formulation are enzymatic stability and preservation against microbial growth. It is sometimes not appreciated that the dominant factor affecting enzyme stability is the intrinsic stability of the enzyme itself; formulation can do very little to correct for a structurally labile protein. Therefore, it is advisable to make stability an important criterion of the initial enzyme screening process.

An important case is the application of enzymes in laundry detergents. Market trends in the United States show that consumers prefer liquids to powder detergents by a ratio of 2 to 1. These products are stored with no temperature control on shelves in the presence of harsh surfactants, such as linear alkylbenzyl sulfonate (LAS) and require extraordinary measures for stabilization. LAS, by its nature is an effective cleaning agent, causes surfactant-induced unfolding in proteins. There are countless examples of the development of stabilization systems in the intellectual property space. A common theme is to reduce the water activity and to use borate/glycol stabilizers that bind to the active site of proteases.

Microbial growth is fairly easy to control by the addition of food grade antimicrobials such as sodium benzoate, methyl paraben, and various other commercial preparations. In addition, reduction of the water activity through use of sequestering agents aids in controlling microbial growth. The prevention of precipitation and hazes is often a highly empirical challenge and one that is very specific to the enzyme and the specific raw materials used in the process. It is best to use ingredients with high solubility, and to screen potential formulations by extended storage at high and low temperatures. In the case of proteins, an additional objective is to keep the pH value as far away as possible from the isoelectric point of the protein, at which the protein is least soluble.

Emulsions are the most commonly used type of formulation in the food, paint, and personal care industries. The typical emulsions are "oil in water" or "water in oil" preparations. Proteins, peptides, metabolites, and other fermentation products can be incorporated into emulsions. Covalent attachment of hydrophobic compounds, such as fatty acids or polyethylene glycols, to proteins or peptides lends these hydrophilic molecules to reside in the lipid phase.

12.4 REGULATORY CONSIDERATIONS

Bioproducts are regulated according to their application, for example, whether enzymes are used in food, feed, detergents, textile processing, and so on. Regulations also differ country by country. Irrespective of legislation, it is prudent for the product manufacturer to ensure that the production process, the potential product, and its intended use are safe prior to introduction. Along with transfer of the strain, fermentation, recovery, and formulation process to the manufacturing plant, validated quality control assays, which are suitable for ensuring product quality, must also be provided. The EPA, FDA, and other agencies throughout the world are all involved in regulation of products derived from naturally occurring organisms and genetically modified organisms (GMO).

Essentially the regulations require:

(1) The use of nonpathogenic organisms,

(2) The absence of endotoxin or other toxins,

(3) A well-documented and reproducible fermentation and downstream process, and

(4) Use of safe and well-characterized organisms and processes.

In response to the latter, not only are the genetically modified organisms well characterized, they are designed to barely survive in the general environment. Because recombinant organisms can be grown under contained good large-scale industry practices, safety issues relating to environmental release are minimized.

In August 2002, the FDA announced a significant new initiative, the Pharmaceutical Current Good Manufacturing Practices (cGMPs-initiative) for the 21st Century, to enhance and modernize the regulation of pharmaceutical manufacturing and product quality. The initiative was intended to modernize the FDA.s regulations for pharmaceutical quality for veterinary and human drugs.

For control and containment of waste, regulatory requirements dictate appropriate containment that is related to the risk presented by the organism. Most of the processes are contained and hence subject to guidelines for industrial applications of recombinant DNA. Containment may be achieved biologically on the basis of inherent properties of the organisms, for example, their survival in the general environment is limited. Also, operation and design of the manufacturing facility minimizes potential releases of the recombinant organism to the environment. The degree of physical containment is matched to the risk presented by the organism. The same considerations apply to products made by recombinant organisms. Air monitoring is performed in production environments to evaluate levels of protein. This is important for assessing the efficiency of containment, respiratory protection requirement, and ensuring that environment levels meet occupational exposure limits.

12.5 REGULATORY REQUIREMENTS

GMP regulations and good business practices require that pharmaceutical raw materials (and also their suppliers) have to be qualified both initially and periodically. Accordingly, manufacturers of API's and finished pharmaceutical dosage forms have to establish the identity, purity and quality of raw materials and excipients with the use of suitable specifications, test methods and acceptance criteria. Similar requirements can be found in the US Code of Federal Regulations, ICH guidance documents, European GMP regulations, and also within ISO.

Raw material testing ensures that the raw materials used in pharmaceutical products are suitable for their intended use. Conducting raw material analysis using appropriate test methods and successfully meeting the acceptance criteria can prevent costly production problems and diminish quality variations.

Patient safety is a key reason for these requirements, dating back to several unfortunate events within the pharmaceutical and food industries, and regulatory requirements in the pharmaceutical industry have evolved over time to reduce the probability, or risk, of such events.

Legally, a pharmaceutical firm takes on full responsibility for the quality of the raw materials it purchases and uses in a GMP manufacturing process. Consequently, it is in the business interest of a firm to exercise reasonable oversight of suppliers and test laboratories, and to characterize raw materials appropriately. Some of the most important actions a firm takes to reduce risk include setting specifications that define and control the raw materials, testing to verify identity and quality, and establishing systems to prevent the use of unsuitable materials.

Raw material qualification should be carefully defined in GMP procedures and placed under strict change control. Raw materials deemed "critical" require testing of more supplier lots for more attributes, and extensive supplier evaluation, before qualification is achieved. The critical status of a raw material is related directly to its intended use in the process and to the potential risk created by a quality deficit in the raw material that may adversely impact the products identity, purity, potency, toxicity, or efficacy. A raw material may be critical to one process but not to another. Each firm must identify which materials are critical and justify the choices made and the additional oversight required.

Manufacturers take various approaches to raw materials testing compliance. Some qualify a raw materials supplier by performing an initial detailed vendor audit followed by an annual qualification consisting of full pharmacopoeial monograph testing on three lots of material. If the qualification lots test successfully, then subsequent materials shipments will require only monograph identification testing. However, companies that take a more conservative approach to raw material release require full monograph testing for each lot of supplied material.

At the moment pharmacopoeial and formulary monographs such as the US Pharmacopoeia-National Formulary (USP-NF), the European Pharmacopoeia and the Japanese Pharmacopoeia provide standardized test methods for the most common and widely used materials. The primary function of pharmacopoeial monographs is to establish minimum standards that set the identity, purity and quality requirements for raw materials.

12.6 MANUFACTURING FORMULAE, PROCESSING AND PACKAGING INSTRUCTIONS

Manufacturing formulae should clearly indicate the product name, potency or strength, and exact batch size. It lists each of the starting raw materials required, and the quantity in which each is required. The processing instructions should contain step-by-step manufacturing instructions. The detail given should be sufficient to allow a technically competent person, unfamiliar with the process, to successfully undertake the manufacturing procedure.

The processing instructions should also indicate the principal items of equipment utilized during manufacture, and the precise location in which each step should be undertaken (e.g. in a specific clean room, etc.). It will also list any specific precautions which must be observed during manufacture (e.g. precautions to protect the product from, perhaps, excessive heating, or to protect the operator from any potentially dangerous product constituent). Each product will also have its own labeling and packing instructions, indicating:

- the label to be used, and its exact text;
- exact packing instructions (e.g. how many units of product per pack, how may packs per shipping carton, etc.).

A copy of the label to be used is generally attached to the documents, to allow the supervisor and operators to verify easily that the correct label has been dispensed for the product in question.

12.7 LABELING AND PACKING

After the product has been filled (and sealed) in its final product container, it is immediately placed under quarantine. QC personnel then remove representative samples of the product and carry out tests to ensure conformance to final product specification. The most important specifications will relate to product potency, sterility and final volume fill, as well as the absence of endotoxin or other potentially toxic substances. Detection and quantification of excipients added will also be undertaken.

Only after QC personnel are satisfied that the product meets these specifications will it be labeled and packed. These operations are highly automated. Labeling, in particular, deserves special attention. Mislabeling of product remains one of the most common reasons for product recall. This can occur relatively easily, particularly if the facility manufactures several different products, or even a single product at several different strengths. Information presented on a label should normally include:

- name and strength/potency of the product;
- specific batch number of the product;
- date of manufacture and expiry date;
- storage conditions required.

Additional information often presented includes the name of the manufacturer, a list of excipients included and a brief summary of the correct mode of product usage. When a batch of product is labeled and packed, and QC personnel are satisfied that labeling and packing are completed to specification, the QC manager will write and sign a Certificate of Analysis. This details the pre-defined product specifications and confirms conformance of the actual batch of product in question to these specifications. At this point, the product, along with its Certificate of Analysis, may be shipped to the customer.

INDEX